AF183272

LEISURE, SPORT AND TOURISM, POLITICS, POLICY AND PLANNING

Leisure, Sport and Tourism, Politics, Policy and Planning

Editor

Regina Belo

scitus academics

Leisure, Sport and Tourism, Politics, Policy and Planning
Edited by **Regina Belo**

ISBN: 978-1-68117-549-2
Library of Congress Control Number: 2016942955

© 2017 by
SCITUS Academics LLC,
www.scitusacademics.com
Box No. 4766, 616 Corporate Way,
Suite 2, Valley Cottage,
NY 10989

This book contains information obtained from highly regarded resources. Copyright for individual articles remains with the authors as indicated. All chapters are distributed under the terms of the Creative Commons Attribution License, which permits unrestricted use, distribution, and reproduction in any medium, provided the original author and source are credited.

Notice

Reasonable efforts have been made to publish reliable data and views articulated in the chapters are those of the individual contributors, and not necessarily those of the editors or publishers. Editors or publishers are not responsible for the accuracy of the information in the published chapters or consequences of their use. The publisher believes no responsibility for any damage or grievance to the persons or property arising out of the use of any materials, instructions, methods or thoughts in the book. The editors and the publisher have attempted to trace the copyright holders of all material reproduced in this publication and apologize to copyright holders if permission has not been obtained. If any copyright holder has not been acknowledged, please write to us so we may rectify.

Printed in United States of America on Acid Free Paper

PREFACE

Leisure travel is travel in which the primary motivation is to take a vacation from everyday life. Leisure travel is often characterized by staying in nice hotels or resorts, relaxing on beaches or in a room, or going on guided tours and experiencing local tourist attractions. Most meals are eaten out when traveling for pleasure, and often more expensive modes of transportation, such as taxis, are used to get around. In some cases, leisure travel might be used to refer to any trip that lasts more than a week, regardless of the primary focus. Leisure travel is generally seen as the opposite of business travel. Leisure travel can be many things, but besides being a vacation, it doesn't have to be any one of them. Although many leisure travelers spend more money than they would in their everyday life, others might choose to travel frugally. The tourism industry has started to recognize sport tourism, i.e. the experience of travel to engage in or view sport-related activities, as an important market. In order to compete in the growing sport tourism market, it is crucial for communities to develop a profound understanding of the benefits and impacts of sport tourism, of the process of bidding for events, possible sponsorship opportunities and other elements involved in the planning and hosting of a successful sport event.

Leisure, Sport and Tourism, Politics, Policy and Planning promote, enhance and disseminate research, good practice and innovation in all aspects of Hospitality, Leisure, Sport and Tourism and Events to its prime audience including teachers, researchers, employers, and policy makers. Encourage greater understanding, links and collaboration across its constituent fields. It covers theoretical perspectives and practical guidelines for the application of a range of analytical techniques, making it valuable tool for students as well as practitioners.

TABLE OF CONTENTS

7

TECHNOLOGY ADOPTION AND THE ACCOUNTABILITY FACTOR: A CASE FOR THE RE-INTERMEDIATION OF TRAVEL AGENCIES IN SMALL ISLAND DEVELOPING STATES

Andrew J. Spencer

Centre for Hotel and Tourism Management, University of the West Indies, Bahamas Tourism Training Centre, Bahamas

ABSTRACT

This paper highlights the importance of trust and accountability to Caribbean consumers of online travel products. It reveals a critical element, which has been largely ignored in the literature related to travel consumer online purchasing decisions. Previous research on factors that influence consumer decisions about online travel purchase tended to focus on so called universal factors such as speed, convenience and cost, to name a few, while those that promote more traditional offline purchasing surround expertise, social interaction, privacy and security. While these cannot be overlooked, this work highlights that an important factor in the decision making process is the ability to hold persons accountable and depend on organizations to see their booking through to a successful end. This is particularly heightened if and when errors in bookings are made. A mixedmethods approach was used in this paper. The process began with 31 in-depth interviews with the Chief Executive Officers of all owner-managed travel agencies in Jamaica. Given that the Caribbean countries bear some similarities, a cross-national survey totaling 302 individuals was conducted between Jamaican and the Bahamas over a three (3) week period. The findings not only reveal that accountability is important to consumers in relationship-oriented societies such as Jamaica and the Bahamas, but that this concept must be explored for other countries and contexts, particularly for those interested in maintaining the buoyancy of travel agencies in a technologically advanced world.

KEYWORDS

accountability, technology adoption, travel agencies in Small Island developing States

1. INTRODUCTION

Travel is a six trillion dollar industry and has become the number one earner of online revenues (Blanke & Chiesa, 2013). Online bookings have become fairly international, with some territories adapting at a faster pace than others. Generally however, the traditional travel supply chain has been distorted in many industries around the world as a result of the World Wide Web. The concept of using Information and Communication Technologies in the travel industry is not a novel one and stems from the first reservation systems in the 1950s to the tourist information systems like Travel Industry Services (TIS) and Gulliver of the 1980s (Neuhofer, Buhalis, & Ladkin, 2012) to major global distribution systems like Semi-automated Business Research Environment (SABRE) and

Amadeus in the 1990s to the enormous number of current travel activities on the web. Now more than ever however, it brings a multiplicity of players together with relatively easy access to each other and each other's information base. The implication of this is that a more level playing field has been created with more options available to the creators of products and services and the final consumer. Internet booking brings new independence for suppliers and travelers (Xiang & Gretzel, 2010). New value chains and value systems are emerging and the players within the industry have to redesign their strategy under the power of Information and Communication Technology.

The international landscape over a decade ago showed internet readiness in some countries. Canadians and North Americans in general are keen users. According to Egger and Buhalis (2011) Europe lags behind North America by 14-18 months for internet penetration. Asian countries such as South Korea and Japan are far ahead of the game. Of particular importance is that some developing countries, particularly those in the Caribbean are experiencing a lag effect. There are obvious implications for countries which are slower on the uptake of ICTs and for businesses operating in those contexts. For example, intermediaries in Jamaica, which are predominantly travel agencies that focus on the outbound traveler, may therefore not experience some of these global issues at the same rate as more developed economies, based on internet readiness of the population.

Any discussion of internet readiness must also take into consideration groups of individuals mentioned by Jensen (2012). He states that some consumers are not computer literate and don't intend to be. She also highlights that some consumers simply prefer human interface whether literate or not, and that some consumers are internet savvy but fear lack of security with credit card transactions. Additionally many experienced internet users still opt to book traditionally for convenience. For many, 'techno-fear' may still exist since the technology changes so frequently. Each technological development brings with it a new "fear" even where changes are marginal. The above mentioned however does not negate the significance of globalization and the need for firms to be competitive in international markets. It is in discussions surrounding the competitiveness of firms, that firms must seek to understand the dynamics of the travel consumer in context.

The above mentioned however does not negate the significance of globalization and the need for firms to be competitive in international markets. It is in discussions surrounding the competitiveness of firms, that firms must seek to understand the dynamics of the travel consumer in context.

2. LITERATURE REVIEW

According to Aldebert, Dang and Longhi (2011), there has been an evolution in travel bookings. Initial processes included traditional travel agents who had access to information stock that was generally unavailable to the consumer market. This then evolved to an environment with traditional e-mediaries, which involved the same travel agencies which now had access to global distribution systems (GDS) such as Amadeus, SABRE and Galileo; all owned by major airlines and gave agencies access to consistent information to avoid duplication of bookings. This was then followed by the advent of the internet which was initially just used for information search and subsequently direct bookings and payment with suppliers. Most recently there has been the emergence of a number of web-based agencies, which refers to a group of travel intermediaries with an online presence.

The use of online services immediately presupposes a level of internet readiness in consumer markets. The World Wide Web is used by millions of people connected to thousands of computers all over the world. The web puts a friendly face on the internet providing an interface that a 10-year old, or even a 70-year old, could easily understand (Middleton & Clarke, 2012). The web defines today's most important activities on the internet which include: communication, collaboration, information gathering, direct advertising and direct online selling

2.1. The Travel Consumer

There is a large body of research which addresses the travel consumer in the internet age. Following on decades of work in the area, Wen (2012) pointed out that convenience, value and merchandise options are the most significant drivers of the consumer's intention to purchase travel products online. Castillo-Manzano and Lopez-Valpuesta (2010) focused less on the attributes of the distribution channel and more on the demographic factors of the consumer, highlighting that the online travel consumer is typically a young, highly-educated, habitual traveler who tends to travel with low cost carriers. Other researchers (Kamarulzaman, 2010; Park, Wang & Fesenmaier, 2011) also support this demographic approach. Arguments which focus solely on demographic characteristics, do not necessarily present an opportunity for struggling firms to find ways of being re-intermediated, but rather identify demographic segments which may now be out of the reach of the traditional agent. Distribution channel studies however, uncover important attributes in pulling consumers to a particular platform and allow players in different modalities to be able appeal to markets based on these variables. Consequently, this paper seeks to focus on travel purchase drivers with a view to identifying gaps in the literature.

Two of the seminal pieces of research in the general area of technology adoption were the diffusion of innovations by Rogers (1962) and the technology acceptance model (Davis, 1989). While the work of Rogers identified more institutional and systemic drivers of adoption; the technology acceptance model (TAM) by Davis emphasized perceived ease of use and perceived usefulness in consumers' decision to use a platform. Andreu, Aldás, Bigné and Mattila (2010) support and extend the TAM in positing that these two areas of perception were critical and that the perceived usefulness played a more pivotal role in organizational contexts.

Peng, Xu and Chen (2013) highlight convenience and social interaction as important to the buying process for the travel consumer, and Tsang, Lai, and Law (2010) add the elements of user friendliness and security to the discussion, as having high importance to consumers. More recently Chen, Hsu and Lin (2010) make the claim that the most important drivers for travel consumers are privacy, safety and product quality. While these elements tend to favour a continuation of offline purchasing Ganguly, Dash, Cyr, and Head (2010) posit that travel consumers place a high value on the level of control given to them by the internet.

Succinctly, attributes such as speed, efficiency, convenience, control and pricing (Ganesh, Reynolds, Luckett, & Pomirleanu, 2010; Jensen, 2012) support online travel purchase. While expertise (Walsh, Hennig-Thurau, Sassenberg, & Bornemann, 2010), social interaction, privacy and security (Nuseir, Arora, Al-Masri, & Gharaibeh, 2010; Nepomuceno, Laroche, Richard & Eggert, 2012) promote traditional purchasing. Therefore, consumer decisions about using online or offline platforms have a direct impact on the travel supply chain and often result in disintermediation.

2.2. Disintermediation

Intermediaries have played a very important role in physical markets as well as in information markets from an economic standpoint. For instance, retailers provide access to goods produced by remote, in many cases overseas, sellers for consumers. They also act to ensure quality and diversity and provide comprehensive product information. Intermediaries are often adding cost for their services to the products that have to be paid by the suppliers and/or consumers. This is typically done through commissions (suppliers) or service charges (consumers).

Essentially, a firm can undertake the same task, but an intermediary was traditionally preferred because it costs less especially for physical markets. It has been argued (Alt & Klein 2011; Doherty & Ellis-Chadwick, 2010; Chircu & Kauffman, 2013) that electronic markets lower the cost of market transactions. It will become easy to directly match buyers and sellers, and as a result the role of intermediaries may be reduced or even eliminated, thus leading to disintermediation. Disintermediation means that products and services are directly given to the consumers, thus having less cost, the web is a possible instrument to support disintermediation by directly connecting customers to producers.

The internet means lower distribution costs, larger market share and higher revenues (Lieber & Syverson, 2011). Supplier companies which are always seeking to cut costs are therefore exploring the option of convenient market access. The Caribbean has experienced some effects of this as American Airlines, British Airways and the regional carrier Air Jamaica cut commission from 9% to 6% for travel agencies in the region in 2000. Kotler (2012) posits that many airlines opt for the internet in the face of inflation and competition because of reduced distribution costs. It must be noted that as a result of technological development, many airlines have become known globally. One of the primary roles of an intermediary was to ensure penetration into geographical markets which are more familiar to these agents because of their location. However, brand awareness has eroded this market ignorance that existed in some regions (Kracht & Wang, 2010).

Reduced transaction cost and commission present a strong case for complete elimination of travel intermediaries (Christodoulidou, Connolly, & Brewer, 2010). When Caribbean travel agencies attempted to boycott the sale of American Airlines tickets in response to the 3% cut in commissions in 2000, their efforts proved futile as the airline, which accounts for approximately 70% of air traffic to the region, simply made it known that "direct bookings" was always an option. Kotler (2012) states that the reduction of commissions paid and the competition provided by the internet present uncertain operating environments. In 2003, British Airways reduced agents' income in the UK to 1%. What seems to be clear from these events is that organizations are in the business of making a profit; either by increasing revenues or reducing costs, or a combination of both. With inflation, competition and rising oil prices, revenue increases in many cases will only be marginal, as cheaper prices may not be passed on to the consumers. Hence the focus is predominantly on reducing distribution cost.

The IATA (2011) framework that covers passenger sales agency agreements is inadequate. As it relates to payments to agencies, it only articulates that agencies shall be paid in "a manner and amount to be determined by airlines." In an environment of major cost cuts, airlines will continue to oppose and reject new passenger sales agreements that seek to include a proper remuneration clause for travel agencies as they did in 2002 in Miami. He argues that agencies are therefore being forced to cut costs and as a result, over two million jobs were lost between 2001 and 2003. In Jamaica the issue of cutting costs has been a real one for travel agencies. In 2003, coping mechanisms were implemented such as the rotation of staff, lay-offs, mergers and tightened inventory. In addition compensation packages had to be changed from fixed salary plus commission to a situation of "pure commission". Changes are further demonstrated in a quote from Pauline's Travel in 2003 which states "some staff are now working from home".

One way for companies to differentiate their products from rival companies is faster and more efficient delivery of products to their customers. The behavior of customers who are already buying goods and services on-line clearly indicates that companies can use internet technology as a digital channel to deliver services to customers (Urban, Sultan, & Qualls, 2000). An important debate however surrounds whether particular markets are very accessible via this route, a missing element in the literature, and this research will attempt to address issues surrounding context and IT uptake.

It is fair to say that some level of disintermediation will take place for certain products, especially those that can be distributed over the internet, easily described or whose asset specificity is low. Intangibles such as travel fall squarely into this category as there is little need for physical outlets for large quantities of items to reach the market, as e-ticketing has become the order of the day. Intermediaries can be defined as market participants who neither supply nor demand but have the aim to facilitate the functioning of the market for which they receive a commission. If development of the internet leads to the simplification of channel structures (Kracht & Wang, 2010) and suppliers choose to deal directly with end-users, then facilitating the functions of the market becomes a role of the supplier and the information stock of the intermediary is eroded.

Until recently travel agencies were the key intermediary with information as their primary stock in trade (Kracht & Wang, 2010). They posit that changes in technology predominantly favour suppliers and many organizations utilize this to reduce distribution costs. As such they state that online travel sales accounts for the largest retail e-commerce activity. Suppliers must then take advantage of cost reduction opportunities as well as the chance to manage the supply process.

The internet does not only mean reduced costs to suppliers. If this were the case then market forces would not enforce disintermediation so intensely. From a demand perspective, consumers also experience reduced transaction costs. There is a clear link between commission cuts and disintermediation. Suppliers aim to reduce distribution costs through reducing commission percentages. This means reduced revenues for travel agencies, who must then reengineer their processes to earn more revenues. One way in which this is done is through the introduction of service charges, which consumers must bear. In many cases these agencies become the most expensive option for consumers who then seek for more affordable ways to purchase travel and tourism services. One option being provided by suppliers is direct online booking, which very directly competes with the less convenient more expensive option of traditional agency bookings.

Casaló, Flavián, and Guinalíu (2010) highlight that travelers are now able to purchase at any time and any place. This is in keeping with the view that modern travelers demand more high quality travel services, products, information and value for money (Chakravarthi & Gopal, 2012). The issue therefore for contemporary consumers is also one of convenience. Xiang and Gretzel (2010) argue that tourists can now receive comprehensive, timely and relevant information in a virtual environment to assist in their decision-making process. The internet provides convenient and instant access for enquiries and bookings at odd times and at a cost advantage (Kimber & Mason, 2013).

It is argued therefore that the internet reduces the importance of travel agencies and may result in them being bypassed (Andreu et al., 2010). This gave rise to the strident position that agencies must change their business practices to survive (Castillo-Manzano & Lopez-Valpuesta, 2010). Xiang and Gretzel (2010) argue that travel agencies must be prepared for restructuring. One example of a restructured, re-focused agency is ThomasCook, which is known to many as Thomascook.com due to its adaptation of online channels. Castillo-Manzano and Lopez-Valpuesta (2010) further posit that there will be a "Natural Selection Process" which will separate those with sound business models and the possibility to expand on all distribution channels and platforms from those who will fail. This therefore suggests that companies must capitalize on online opportunities for distribution. They should not treat the internet as a threat but rather take advantage to offer additional valueadded service that was once not possible (Huang, Yung, & Yang, 2011). Kracht and Wang (2010) argue that although somewhat a threat, the World Wide Web has created opportunities for flexibility and innovative agencies have developed online strategies to supplement offline channels.

Grimm, Hofstetter, and Sarkis (2012) argue that there is a new independence which is being experienced by supplier as well as consumer. The internet offers an effective means of developing a single and sustainable electronic infrastructure for information gathering and business transactions for both travelers and suppliers (Granados, Gupta, & Kauffman, 2010). They argue that a natural outcome is one to one marketing and mass customization. They further state that the contemporary consumer, who requires higher quality, has a greater propensity to purchase online with increasing income and education. Age also plays a significant role for internet usage with the majority of internet users being between the ages 18-34 (File, 2013).

The implications are clearly greater in large developed countries with a higher per capita income, greater access to education and overall higher Human Development Index. The suggestion therefore is that small developing states typically grasp new opportunities such as information technology more slowly. This may explain why the Caribbean in particular has been slower on the uptake of ICTs than the United States and Europe. A threat such as disintermediation may therefore be an issue in the longer term in such regions especially in markets such as Jamaica where the major market is outbound and therefore comprises residents of that country. However the implications are greater for other local tourism suppliers as inbound markets influence most of their business.

Many argue that discussions surrounding disintermediation are premature as what appears to be a likely challenge for agencies may just have presented greater opportunities. Although somewhat a threat, the World Wide Web has created opportunities for flexibility and innovative agencies have developed online strategies to supplement offline channels (Ashari, Heidari, & Parvaresh, 2014). The orientation of an intermediary toward the web will therefore determine its ability to maximize on its potential benefits. They therefore highlight that travel consultants do not seem to worry as they feel they are still deemed important and that they will remain secure if their advice-offering capability were strengthened on the internet rather than simply being booking agents. The argument of strengthening their capabilities makes the important suggestion that with current weaknesses there is cause for concern and that changes will be necessary.

Others (Busby & Huang, 2012) posit that the internet does not change the role of intermediaries who exist to simplify the buyer's choice process. They argue that making choices may have become just as confusing with the magnitude of information and options available to consumers. With greater information about international players being more available and accessible, and with new online entrants to markets, choosing may be difficult for consumers. The most significant role of the intermediary is to reconcile the different needs of producers and consumers. Producers or suppliers typically aim to provide a large quantity of a limited range of goods and services. Consumers on the other hand tend to demand a relatively small quantity of a wide range of goods and services. Intermediaries must then be able to bring these together in one place through an understanding of both needs. The heterogeneous nature of consumers' travel needs makes it important for intermediaries to reconcile the differences by retailing goods and services and meeting the needs of supplier and end-user. It is here that the need for the new info-mediary who gathers information on both parties and merge interests becomes vital.

Granados, Gupta and Kauffman (2012) highlight a number of factors which may contribute to the longevity of the traditional offline agency. They argue that the inability to locate websites, time factors, 'data overwhelm', lack of internet security and lack of human interface will ensure agency existence. Another point argued here is that they will continue to be indispensable for more complex travel arrangements. There is still some difficulty finding websites especially if there is relatively little experience using search engines. Phelan, Christodoulidou, Countryman, and Kistner (2011) support their position by pointing out weaknesses in some airline web pages such as the length of time for loading and the length of time for completing online reservations based on the number of pages. They also argue that there are some general distribution problems on the net for airlines. In addition Almunawar, Anshari, and Susanto (2013) categorize some groups of travel consumers stating that some are not computer literate and don't intend to be. Others simply prefer human interface even though they may be computer literate. Many other consumers they say are savvy but have great concerns about the perceived lack of security with using credit cards online. And so even many experienced internet users still opt to book traditionally. They posit that the issue of "techno fear" is real one which is not likely to be eroded soon, especially since the technology changes so rapidly. Galliers and Leidner (2014) argue however that as buyers' initial curiosity with the web wanes and subsidies end, companies offering services online will be forced to demonstrate that they provide real benefits.

Many discussions surround the statement that travel agents allow for better delivery of service. They save travelers a lot of time when planning and arranging travel, forge relationships and provide unbiased advice which adds value for consumers (Offutt, 2013). He purports that travelers use the internet for less expensive, less complicated travel to more familiar destinations. This is therefore a niche market for travel agencies focusing on specific, exotic destinations. Additionally he mentions business travelers who are mandated to take trips and whose arrangements are handled by travel agencies due to long-standing business relationships and in some cases the complexity of travel and accommodation arrangements. He argues that the service provided by travel agents is superior because of quality assurance, reliability, security and the ability to handle complexity. Travel agents should therefore utilize the internet to improve the quality of advice by providing a synthesis of available data through converting large masses of data into meaningful information for consumers.

The above argument made by Almunawar et al. (2013) identifies what many may consider to be real categorizations. It has failed however to take into consideration the changing landscape. While the baby boomers still account for a large percentage of market share for travel and tourism services, generation x and generation y also include fairly active travelers, who will be the major income earners of the future (Kotler, 2012). It must be noted that these groups are much more fluent with computer, digital and information technology and are value-seekers who search for new and innovative products constantly. The implication therefore is that consumers who have no interest in becoming computer literate are a "dying breed".

A consistent theme in most of the literature seems to be that it cannot be business as usual. According to Assaf, Barros, and Machado (2011) travel agents will always exist, however their methods and functions will change. The ability of traditional agencies to adapt directly impacts how suppliers may do business in the future. Tsang et al. (2010), state that innovative and successful agencies are doing more to provide comprehensive information and additional services. They are responding quickly to customer requests and building relationships with customers by tracking and targeting website users. It is those that are able to change with the landscape that will survive. A multi-platform approach that utilizes various available online and offline channels to serve different users at different times will enable the survival of the fittest (Xiang, Pan, Law, & Fesenmaier, 2010). It is these opportunities that provide the foundation for Smithson, Devece, and Lapiedra (2011) to state that talk of disintermediation has been premature and that the reverse process of reintermediation is more likely.

2.3. Re-Intermediation

It must be noted that the web also provides new possibilities and technologies for companies to create electronic marketplaces that support new types of intermediaries, a process known as reintermediation. Many labels have been used to identify an intermediary in the new dispensation such as: info-mediary, e-mediary, cyber-mediary, e-tailors and virtual intermediaries. These new intermediaries perform functions that include matching buyers and sellers providing product information to buyers and marketing information to sellers, aggregating information goods, providing trust, and integrating components of consumer processes. It must be noted that while most of the names given to the "new" agency suggest extensive, almost exclusive use of cyber space, the term 'info-mediary" puts into perspective the contemporary role which must be performed, although it cannot be ignored that a major vehicle for reintroducing agencies into the supply chain will be cyberspace.

Bronner and de Hoog (2010) argues that the independent travel agent has a future provided it makes appropriate investment in technology, works closely with niche tour operators and creates competitive differentiation by focusing on specialist markets where core assets such as staff expertise can add value to the buying process and prevent the customer from 'going it alone'. The fundamental position of their work is that the internet is complementary rather than a substitution for existing agencies.

The info-mediary will take the new role of brokering relationships between consumers and producers in the world of e-commerce. In addition to having an online presence, agents will still act as counselors, therefore they should be able to access and process large amounts of information to narrow down choices and match with consumer preferences (Kracht & Wang, 2010). Often the users get so much information that they cannot use it in a meaningful way. It is the job of intermediaries to eliminate these problems by providing a platform for information exchange between buyers and sellers, which aggregates the relevant information and brings the appropriate trading partners together.

Tsang et al. (2010) posit that the onslaught of many web-based agencies such as Expedia, ebookers and Travelocity signals the need for the middleman in the virtual environment. The process of restructuring in the travel and tourism industry has been driven by the rapid growth of 'emediaries' (Nalazek, 2011). Every market has some main components: market participants (supplier, buyer, wholesaler, and retailer), products and services, and transactions and processes (Palladino, 2010). The market is the location where the supplier and customer meet and whose activities are governed by price and competition. These components of the market may exist physically or digitally. A significant problem in respect of markets is the information deficiency or asymmetry between the market participants. This allows intermediaries to position themselves to solve these problems. The advent of the internet has made information very available for those suppliers operating on online platforms and so the issue of information deficiency becomes less of a problem. Information asymmetry however still exists to a large extent, in fact it may be more so. According to Tanford, Baloglu, and Erdem (2012), making choices may have become just as confusing for individuals trying to navigate a path and decide on the best option

It may be argued that absence from digital economies and digitalized markets is uncompetitive in a globalized borderless environment where rivals capitalize on opportunities. Travel intermediaries therefore have the option of adding value online in much the same way that value has traditionally been added offline. This may however require more detailed analyses and business strategies that address what needs to be added as the needs of the online, digitalized market may vary somewhat from the traditional, physical, offline markets.

Other agencies have been re-intermediated through strengthening relationships. Travel agent frequent–users are more involved, enthusiastic, innovative, and take more trips (Mak, 2011). As a result of being more frequent flyers, they continuously gather more travel information. Some agencies have therefore identified heavy users, discovered their wants, needs and feelings and developed ties through guaranteed satisfaction and individualized attention. It must be noted that the minority of consumers account for the largest share of travel sales for agencies, somewhat in conformity to the Pareto Principle. The information technology also aids these efforts through customer relationship management programmes (Egger & Buhalis, 2011). Additionally, relationships are also more easily formed in smaller markets and in contexts that are more culturally predisposed to relationship building. This brings to the fore the needs of consumers and points to discourse surrounding what attributes are most important to them in the booking process.

3. METHODOLOGY

3.1. Research Question

What are the key factors that contribute to the differences in the rate of disintermediation between Caribbean travel firms and Developed Countries?

3.2. Objectives

1) To identify moderating consumer variables that influence the decision to book online in the context of Jamaica and the Bahamas.

2) 2. To explore the views of managers in Jamaican travel firms about their ability to compete in a new technologically driven landscape.

To address the research question, a mixed-methods approach was used in this paper. The process began with 31 in-depth interviews with the Chief Executive Officers of all owner-managed travel agencies in Jamaica. Given the small number of firms now existing in Jamaica, it was decided that the entire population of 31 firms should be investigated, rather than using a sample for the qualitative section of the study. The agencies are "brick and mortar" type firms and the interviews revealed that in addition to previously identified variables such as social interaction, security, efficiency and price, the local market being served wanted "someone to blame" if there was a booking error; hereafter referred to as the "accountability factor".

This recurring theme created a need to gather data from actual and potential travel consumers. Given that the Caribbean countries bear some similarities, a cross-national survey was conducted between Jamaica and the Bahamas over a three (3) week period to extend the study to another destination with similar mass tourism products. The sampling was purposive in that it aimed at getting a majority of young potential and actual travel consumers below the age of 35, as they represent the group, which most likely to engage in online purchases in the literature. By extension this means they are the group least likely to remain loyal to traditional agencies. The sample consisted of 92 males and 210 females, totaling 302 individuals. There were 211 Jamaicans and 88 Bahamians in the study. Approximately 61% of the sample were students and 35% employed. The questionnaire consisted of 23 closed-ended questions on demographics and comparisons between online booking and travel agencies. There were three (3) computed dependent variables: trust, preferences and accountability.

The instrument was designed based on the questionnaire adopted from the work of Wöber and Gretzel (2000). However many items had to be altered and created as there was a gap in the existing literature regarding trust and accountability as factors for consumer online decisions. Trust was computed from questions on: whether close friends trust travel agencies (q18), feeling safer with a travel agency (q19), belief that the elderly are safer with a travel agency (q20) and recommending first time travelers to travel agencies (q21). Preference was formulated from: family members preferring online booking (q14) and whether close friends prefer online booking. The accountability factor was derived from: belief that it is good to have access to a travel representative (q22) and belief that travel agents have a greater responsibility in booking than an online booking customer system (q23). Finally, basic analysis was done using chi-statistic to establish relationships and Spearman's correlation, Mann-Whitney and Kruskal Wallis were applied to further test these statistical relationships. The results were then compared to the qualitative interviews with CEOs to assess consistency of findings at each stage of the data analysis process.

4. FINDINGS AND DISCUSSION

The findings from the primary data collection reveal an element which is largely ignored in the literature on travel consumer online purchasing decisions: the issue of accountability. The qualitative interviews highlighted that traditional agents feel that they may be able to stave off the threat of becoming irrelevant in a relationship-oriented Jamaican society. Some respondents articulated that:

"Our customers are trying to book for themselves online but always come back to us because one wrong click can be disastrous. So when they make mistakes they always come for us to fix it when it is too late. I always use it as a teaching opportunity so that the next time they will just come to me first."

"The people who come to me are afraid of the credit card security issue with the internet but also they don't have the confidence that they will get the booking correct so they want somebody to blame if it goes wrong. Also if it gets messed up it would not be an additional cost to them to fix it from our end."

"There is a technology fear in our society to the point that when some of my clients who have tried and failed at using some of these technologies to do business, find a good deal online, they still come to me asking me to use the company credit card to do the booking. Needless to say I also charge them for this service."

This recurrent theme in the responses from the elite interviews prompted a need to collect data directly from the consumer. The quantitative analysis revealed the following

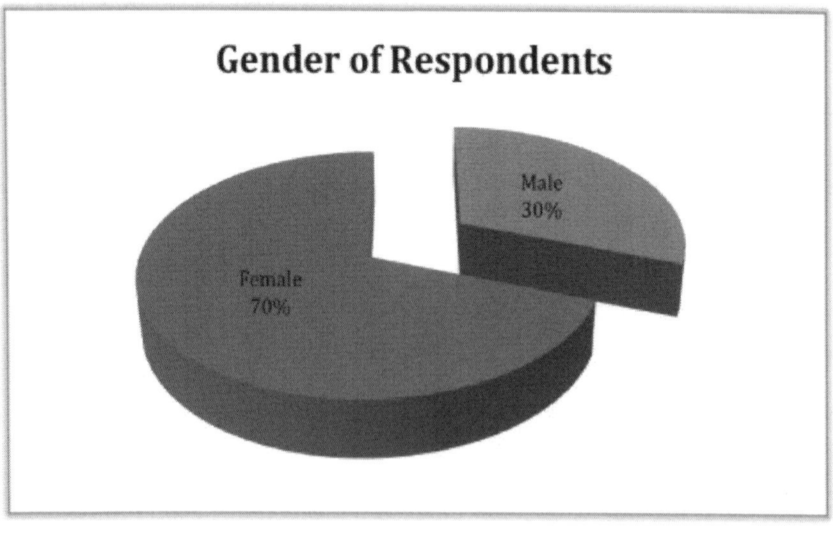

Figure 1. Showing gender distribution of respondents

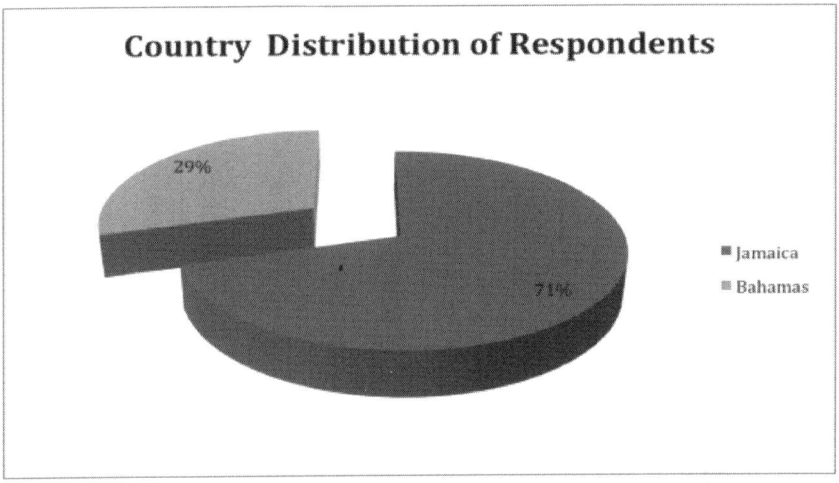

Figure 2. Showing country distribution of respondents

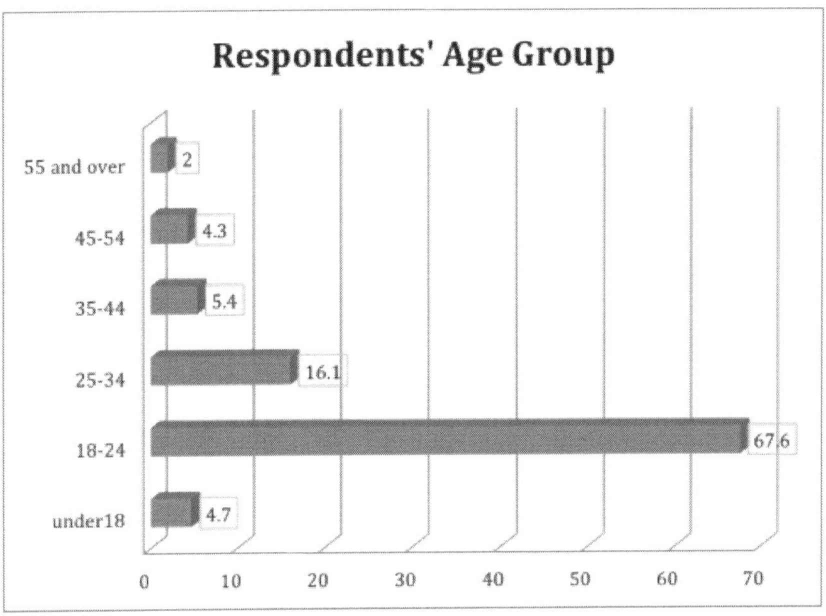

Figure 3. Showing respondents' age group

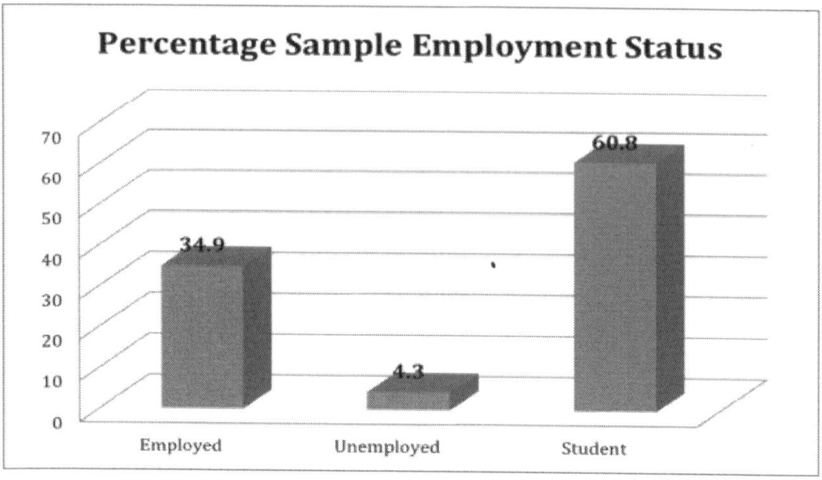

Figure 4. Showing percentage employment status

The population size of Jamaica (approx. 2.7 mil.) warranted having a larger sample than the Bahamas (approx. 300,000) in an effort at being proportionate.

Figures 1, 2, 3 and 4 indicate the demographic overview of the respondents. The convenience sample was a young population of generation y (born 1980s and 1990s) travelers who were predominantly students. This was deliberate in that the market share which is more likely to be lost by traditional agencies is the young consumer who is more internet savvy (Castillo-Manzano & Lopez-Valpuesta, 2010). There is value in understanding their preferences in light of the prevailing argument in the literature that online bookings are largely driven by younger consumers with less loyalty to travel agencies, who will be the major income earners of the future (Kotler, 2012).

4.1. Bivariate Analysis

The chi-statistic reported a statistically significant relationship between levels of trust and country of residence ($\chi2=10.004$, (3), .181) as indicated in table 1A. This relationship is very weak and only explains approximately 3.2% of the variation observed in trust. Notably on a whole, 10% of the sample had no trust in travel agencies while approximately 20% showed high levels of trust. Further disaggregation reveals that 26.4% of Bahamians compared to 17.1% Jamaicans have high levels of trust in travel agencies. More than two-thirds of the Bahamian sample indicated mid to high levels of trust while less than 50% of the Jamaican sample indicated similar trust levels. Table 1B below goes further by providing analysis into the differences between the countries regarding trust levels and travel agencies.

Table 1A. The relationship between trust and country of residence

Level of Trust	Country of Residence		
	Jamaica n=210	Bahamas n=87	Total n=297
None	12.4	5.7	10.4
Low	40.5	27.6	36.7
Mid	30.0	40.2	33.0
High	17.1	26.4	19.9

P<.05

Table1B. The mean difference in trust by country of residence

Country of Residence	N=297	Mean Rank
Jamaica	210	139.58
Bahamas	87	171.73

U=7157.500, p=.002, .175

There is a very weak significant difference between Jamaicans and Bahamians in trust levels in travel agencies (U=7157.7500, p=.002, .175). Bahamians have higher trust levels and this is evidenced with a mean rank of 171.73 compared to 139.58. This difference explains approximately 3.1% of the differences observed between Jamaica and the Bahamas.

Table 2. The relationship between trust and travel agency usage intent

Level of Trust	Do you intend to use a travel agency next travel time		
	Yes n=103	No n=182	Total n=285
None	3.9	12.6	9.5
Low	24.3	44.5	37.2
Mid	40.8	29.1	33.3
High	31.1	13.7	20.0

$P<.05$

As Table 2 shows, there is a significant relationship between trust and intent to use a travel agency for the next trip ($\chi2$ = 25.121 (3) .285). This relationship is weak and explains 8.1% of the variation observed within levels of trust. Approximately 43% of the sample who reported that next travel time they would not be using a travel agency also indicated mid to high levels of trust in travel agencies. This is counter intuitive and may be explained by the fact that younger consumers will typically be driven by convenience (Wen, 2012). It must however be highlighted that this does not constitute a majority of the sample. This can be compared to 28.2% of those indicating they would use a travel agency next travel time but "low to no" levels of trust. The implication is that higher levels of trust increase the likelihood of future bookings, which is intuitively aligned.

Table 3. The relationship between trust and travel agency usage

Level of Trust	Do you use travel agencies		
	Yes n=99	No n=137	Total n=236
None	0.0	16.1	9.3
Low	23.2	43.1	34.7
Mid	45.5	27.7	35.2
High	31.3	13.1	20.8

$P<.05$

Levels of trust in travel agencies and travel agency usage have a statistically significant relationship ($\chi2$= 36.676 (3) .367) as indicated in table 3. This is a weak relationship and explains approximately 13.5% of the observed variation in levels of trust. Notably 31.3% of those persons who indicated that they use travel agencies also highlighted that they have a high level of trust in the agencies compared to 13.1% of those who said they do not use but have similar trust levels. In fact, approximately 41% of those persons who do not use travel agencies indicate mid to high levels of trust. This is non-intuitive and points to the influence of other factors explored earlier in the literature such as speed, efficiency, convenience, control and pricing (Ganesh et al., 2010; Jensen, 2012). Of importance is the fact that trust and accountability are not expected to explain why all consumers do not book with travel agencies but rather to add an element that was not previously considered in a small island state context. On the other hand 23.2% of those that actually do use travel agencies have low levels of trust.

Table 4. The relationship between levels of trust and travelling experience

Level of Trust	Travelling Experience		
	Yes n=227	No n=71	Total n=298
None	9.3	14.1	10.4
Low	34.8	43.7	36.9
Mid	36.1	23.9	33.2
High	19.8	18.3	19.5

$p>.05$

Table 4 illustrates that there is no relationship between levels of trust with travel agencies and travel experience1 ($\chi2 = 4.884$ (3), $p>.05$). Those persons who have never travelled have a slightly higher distrust with approximately 58% reporting none to low levels of distrust compared to 44% of those with travelling experience. Notably there is a difference but this is only 14%. Therefore travel experience does not significantly influence trust in travel agencies. This is interesting and nonintuitive as it would be expected that perceptions would be influenced by experience. This highlights an even greater need for agencies to be expert advisors and provide a service which generates excellent word of mouth for the potential traveler who has never previously booked.

Table 5. The correlation between trust and preferences

	Trust	Preferences
Trust	1.00	.019
Preferences	.019	1.00
Correlation	.137	...
N	300	295

P<.05

There is significant weak positive association between trust and preferences seen in table 5 (p=.019, .137). Approximately 2% of the variation in preferences can be explained by trust. Of particular importance however is the importance of holding someone accountable for bookings and possible errors.

Table 6. The accountability factor and country of residence

Accountability	Country of Residence		
	Jamaica n=208	Bahamas n=88	Total n=296
Unimportant	4.8	0.0	3.4
Important	95.2	100	96.6

P<.05 (Fisher's exact)

Accountability is determined through responses which indicate that consumers believe it is good to have access to a travel representative, especially in terms of error reconciliation (q22), and the belief that travel agents have a greater responsibility in booking than an online booking customer system (q23). Table 6 indicates that a total of 96.6% of respondents between both countries are more inclined to book with a travel agent on the basis of accountability. In terms of differences between countries of residence, there is a weak significant relationship ($\chi2$ =4.379, p=.036, .122). This is an extremely interesting finding as it points to possible similarities throughout the Caribbean region, which may render traditional agencies useful, and presents a need to replicate studies of this nature in other small island developing states. Approximately 2% of the variation in accountability can be explained by country of residence. Due to the cell count a fisher's exact was used as is the norm for contingency tables with small sample sizes.

5. CONCLUSION

Previous research on factors that influence consumer decisions about online travel purchase tended to identify factors that were considered to be universal. Typically online purchasing is said to be related to speed, efficiency, convenience, control and cost, while those that promote more traditional offline purchasing surround expertise, social interaction, privacy and security.

A key factor, which suggests a reason for why some consumers prefer offline booking channels (travel agents or direct), has been ignored and this may be applied to contexts in which "technofear" is high and where members of the population do not have extensive experience engaging in ecommerce. As indicated by the data, accountability is important to consumers in relationshiporiented societies such as Jamaica and the Bahamas where there is still a fair degree of uncertainty about online purchasing in general and in particular for a high-risk intangible service such as travel.

Qualitative results from CEOs stated that outbound consumers from Jamaica were less inclined to book online for fear of making errors and wanting someone to blame in the event the booking is incorrect. This was supported by the quantitative data in both countries and accountability was seen as a significant inhibitor to booking online. A combination of factors more clearly identify reasons for the lag in online booking uptake in these countries based on the data presented earlier. These are trust, family preferences, previous usage and accountability. These more adequately explain technology acceptance and adoption in these cases. It must be noted however, that the results of this study are generalizable only to small island developing states with similar characteristics. Future research may assess the impact of accountability in larger developing states as well as developed countries.

REFERENCES

1. Aldebert, B., Dang, R. J., & Longhi, C. (2011). Innovation in the tourism industry: The case of Tourism@. Tourism Management, 32(5), 1204-1213.

2. Almunawar, M. N., Anshari, M., & Susanto, H. (2013). Crafting strategies for sustainability: How travel agents should react in facing a disintermediation. Operational Research, 13(3), 317-342.

3. Alt, R., & Klein, S. (2011). Twenty years of electronic markets research—Looking backwards towards the future. Electronic Markets, 21(1), 41-51.

4. Andreu, L., Aldás, J., Bigné, J. E., & Mattila, A. S. (2010). An analysis of e-business adoption and its impact on relational quality in travel agency–supplier relationships. Tourism Management, 31(6), 777-787.

5. Ashari, H. A., Heidari, M., & Parvaresh, S. (2014). Improving SMTEs' business performance through strategic use of information communication technology: ICT and tourism challenges and opportunities. International Journal of Academic Research in Accounting, Finance and Management Sciences, 4(3), 1-20.

6. Assaf, A. G., Barros, C. P., & Machado, L. P. (2011). The future outlook for Portuguese travel agents. Tourism Economics, 17(2), 405-423.

7. Blanke, J., & Chiesa, T. (Eds.). (2013). The travel & tourism competitiveness report 2013: Reducing barriers to economic growth and job creation. Geneva: World Economic Forum. Retrieved from http://www3.weforum.org/docs/WEF_TT_Competitiveness_Report_2013.pdf.

8. Bronner, F., & de Hoog, R. (2010). Consumer-generated versus marketer-generated websites in consumer decision making. International Journal of Market Research, 52(2), 231-248.

9. Busby, G., & Huang, R. (2012). Integration, intermediation and tourism higher education: Conceptual understanding in the curriculum. Tourism Management, 33(1), 108-115.

10. Casaló, L. V., Flavián, C., & Guinalíu, M. (2010). Determinants of the intention to participate in firm-hosted online travel communities and effects on consumer behavioral intentions. Tourism Management, 31(6), 898-911.

11. Castillo-Manzano, J. I., & López-Valpuesta, L. (2010). The decline of the traditional travel agent model. Transportation Research Part E: Logistics and Transportation Review, 46(5), 639-649.

12. Chakravarthi, J. S. K., & Gopal, V. (2012). Comparison of traditional and online travel services: A concept note. IUP Journal of Business Strategy, 9(1), 45-58.

13. Chen, Y. H., Hsu, I. C., & Lin, C. C. (2010). Website attributes that increase consumer purchase intention: A conjoint analysis. Journal of Business Research, 63(9-10), 1007-1014.

14. Chircu, A. N., & Kauffman, R. J. (2013). Digital intermediation in electronic commerce—The eBay model. In S. Barnes & B. Hunt (Eds.), E-commerce and v-business: Business models for global success (pp.45-66). Taylor & Francis.

15. Christodoulidou, N., Connolly, D. J., & Brewer, P. (2010). An examination of the transactional relationship between online travel agencies, travel meta sites, and suppliers. International Journal of Contemporary Hospitality Management, 22(7), 1048-1062.

16. Davis, F. D. (1989). Perceived usefulness, perceived ease of use, and user acceptance of information technology. MIS Quarterly, 13(3), 319-340. Andrew J. Spencer Submitted on Augu

17. Doherty, N. F., & Ellis-Chadwick, F. (2010). Internet retailing: The past, the present and the future. International Journal of Retail & Distribution Management, 38(11/12), 943-965.

18. Egger, R., & Buhalis, D. (Eds.). (2011). Etourism case studies. Routledge.

19. File, T. (2013). Computer and internet use in the United States. Current Population Survey Reports, P20-568. Washington, DC.: US Census Bureau. Retrieved from https://www.census.gov/prod/2013pubs/p20-569.pdf.

20. Galliers, R. D., & Leidner, D. E. (Eds.). (2014). Strategic information management: Challenges and strategies in managing information systems (revised ed.). Routledge.

21. Ganesh, J., Reynolds, K. E., Luckett, M., & Pomirleanu, N. (2010). Online shopper motivations, and e-store attributes: An examination of online patronage behavior and shopper typologies. Journal of Retailing, 86(1), 106-115.

22. Ganguly, B., Dash, S. B., Cyr, D., & Head, M. (2010). The effects of website design on purchase intention in online shopping: the mediating role of trust and the moderating role of culture. International Journal of Electronic Business, 8(4/5), 302-330.

23. Granados, N., Gupta, A., & Kauffman, R. J. (2010). Research commentary-information transparency in business-to-consumer markets: Concepts, framework, and research agenda. Information Systems Research, 21(2), 207-226.

24. Granados, N., Gupta, A., & Kauffman, R. J. (2012). Online and offline demand and price elasticities: Evidence from the air travel industry. Information Systems Research, 23(1), 164- 181.

25. Grimm, J. H., Hofstetter, J. S., & Sarkis, J. (2012). Understanding diffusion of corporate sustainability standards through sub-supplier management in the food supply chain (WORKING PAPER NO. 2012-25). Worcester, MA: George Perkins Marsh Institute, Clark University. Retrieved from https://www.clarku.edu/departments/marsh/news/WP2012-25.pdf.

26. Huang, L., Yung, C. Y., & Yang, E. (2011). How do travel agencies obtain a competitive advantage?: Through a travel blog marketing channel. Journal of Vacation Marketing, 17(2), 139-149.

27. International Air Transport Association (IATA). (2011). Travel agent's handbook — Resolution 814 edition (effective 1 June 2011). Montreal [etc.]: IATA.

28. [28] Jensen, J. M. (2012). Shopping orientation and online travel shopping: The role of travel experience. International Journal of Tourism Research, 14(1), 56-70.

29. Kamarulzaman, Y. (2010) Geodemographics of travel e-shoppers: An empirical analysis of UK consumers. European Journal of Social Sciences, 16(2), 195-205.

30. Kimber, J. D., & Mason, C. M. (2013). Internet use by small business: An exploratory study in regional Australia. Australia: CSIRO.

31. Kotler, P. (2012). Kotler on marketing: How to create, win, and dominate markets. Simon and Schuster.

32. Kracht, J., & Wang, Y. (2010). Examining the tourism distribution channel: Evolution and transformation. International Journal of Contemporary Hospitality Management, 22(5), 736- 757.

33. Lieber, E., & Syverson, C. (2011). Online vs. offline competition. In M. Peitz & J. Waldfogel (Eds.), The Oxford handbook of the digital economy. Oxford: Oxford University Press.

34. Mak, B. L. M. (2011). ISO certification in the tour operator sector. International Journal of Contemporary Hospitality Management, 23(1), 115-130.

35. Middleton, V. T. C., & Clarke, J. R. (2012). Marketing in travel and tourism (3rd ed.). Routledge.

36. Nalazek, M. (2011). Orbis on line: Internet distribution strategies for a regional hotel group. In R. Egger & D. Buhalis (Eds.), eTourism: Case studies (pp. 74–82). Routledge.

37. Nepomuceno, M. V., Laroche, M., Richard, M. O., & Eggert, A. (2012). Relationship between intangibility and perceived risk: moderating effect of privacy, system security and general security concerns. Journal of Consumer Marketing, 29(3), 176-189.

38. Neuhofer, B., Buhalis, D., & Ladkin, A. (2012). Conceptualising technology enhanced destination experiences. Journal of Destination Marketing & Management, 1(1-2), 36-46.

39. Nuseir, M. T., Arora, N., Al-Masri, M. M. A., & Gharaibeh, M. (2010). Evidence of online shopping: A consumer perspective. International Review of Business Research Papers, 6(5), 90-106.

40. Offutt, B. (2013). Travel innovation & technology trends: 2013 and beyond. PhoCusWright White Paper. New York, NY: PhoCusWright Inc. Retrieved from http://www.sabrepacific.com.au/extra/2013-WORK/131006-Whitepaper-Travel-InnovationTechnology-Trends-2013.pdf.

41. Palladino, A. P. (2010). Zara and Benetton: Comparison of two business models (Master thesis). Barcelona, Spain: Organización de Empresas, Universitat Politècnica de Catalunya. Retrieved from http://upcommons.upc.edu/handle/2099.1/9620.

42. Park, S., Wang, D., & Fesenmaier, D. (2011). Assessing structure in American online purchase of travel products. Anatolia: An International Journal of Tourism and Hospitality Research, 22(3), 401-417.

43. Peng, H., Xu, X., & Chen, W. (2013). Tourist behaviors in online booking: A new research agenda. Communications in Information Science and Management Engineering, 3(6), 280- 285.

44. Phelan, K. V., Christodoulidou, N., Countryman, C. C., & Kistner, L. J. (2011). To book or not to book: The role of hotel web site heuristics. Journal of Services Marketing, 25(2), 134-148.

45. Rogers, E. M. (1962). Diffusion of innovations. New York: Free Press of Glencoe.

46. Smithson, S., Devece, C. A., & Lapiedra, R. (2011). Online visibility as a source of competitive advantage for small- and medium-sized tourism accommodation enterprises. Service Industries Journal, 31(10), 1573-1587.

47. Tanford, S., Baloglu, S., & Erdem, M. (2012). Travel packaging on the internet: The impact of pricing information and perceived value on consumer choice. Journal of Travel Research, 51(1), 68-80.

48. Tsang, N. K. F., Lai, M. T. H., & Law, R. (2010). Measuring e-service quality for online travel agencies. Journal of Travel & Tourism Marketing, 27(3), 306-323.

49. Urban, G. L., Sultan, F., & Qualls, W. J. (2000). Placing trust at the center of your internet strategy. MIT Sloan Management Review, 42(1), 39-48.

50. Walsh, G., Hennig-Thurau, T., Sassenberg, K., & Bornemann, D. (2010). Does relationship quality matter in e-services? A comparison of online and offline retailing. Journal of Retailing and Consumer Services, 17(2), 130-142.

51. Wen, I. (2012). An empirical study of an online travel purchase intention model. Journal of Travel & Tourism Marketing, 29(1), 18-39.

52. Wöber, K., & Gretzel, U. (2000). Tourism managers' adoption of marketing decision support systems. Journal of Travel Research, 39(2), 172-181.

53. Xiang, Z., & Gretzel, U. (2010). Role of social media in online travel information search. Tourism management, 31(2), 179-188.

54. Xiang, Z., Pan, B., Law, R., & Fesenmaier, D. R. (2010). Assessing the visibility of destination marketing organizations in Google: A case study of convention and visitor bureau websites in the United States. Journal of Travel & Tourism Marketing, 27(7), 694-707.

2

BRAND EQUITY AND ITS ELEMENTS: CASE OF THE LAKE BALATON (HUNGARY)

Sulyok Judit1 , and Gábor Michalkó2

[1]University of Pannonia, Hungary.

[2]Geographical Institute, Research Centre for Astronomy and Earth Sciences, Hungarian Academy of Sciences,
Budapest Corvinus University of Budapest, Hungary

ABSTRACT

The article provides a valuable insight into understanding brand equity and its elements that endows the destination with a unique character. Using empirical data, the study seeks to identify potential brand equity items of a destination type, namely waterside area (Lake Balaton, Hungary). The innovative approach of the research is its holistic view, because it integrates the customer-based brand equity with stakeholders' (including media and tourism professionals) perceptions. This enables a complex understanding of the researched topic, the conclusions are summarized in the Five-Stage Brand Pyramid model. The results highlight four main dimension of the brand equity: (1) the fundamental role of destination specific (waterside in this case) attributes, (2) the country of origin elements (emotional and rational benefits of domestic travel in this case), (3) the tourism products/activities, and (4) the emotional dimension. Furthermore, the research also identifies some important gap between demand and supply side. The research resulted important theoretical (structure of brand equity, need for a complex methodology) and practical (strong emotional perceptions of consumers, comprehensive analysis of tourism products/activities) implications that can be scope of further researches.

KEYWORDS

tourism, branding, image, waterside destination, Lake Balaton, Hungary

1. INTRODUCTION

Due to the increased competition among tourism destinations (coastal areas, mountain areas, cities, rural spaces etc.), branding comes more and more into the focus of stakeholders, including theoretical and empirical researches. It is widely accepted that destination brand is a powerful tool for differentiation (Tasci, Gartner, & Cavusgil, 2007; Boo, Busser, & Baloglu, 2009; UNWTO & ETC, 2009). Although every destination offers a unique travel experience, the typologising of tourism receiving areas is a well-known phenomenon for researchers (e.g. Buhalis, 2000).

Among the numerous places to visit all over the world, waterside destinations play the leading role (UNWTO, 2011); coastal tourism is one of the sector's fastest growing area (Hall, 2001). Dating back to the 19th century (Agarwal, 1997 & 1999), coastal destinations reached their peak in the second half of the 20th century. After the long decades of mass tourism era, today travellers are more active, take more trips with a shorter length of stay during which the time spent with family and friends has an important value (Yeoman, 2007). These trends affected fundamentally the waterside areas offering the experience of rest and relaxation (3S/4S segment) for a very long time. This affects also the development potential of high quality/niche products (e.g. wellness is considered a high quality/value tourism product (Konu, Tuohino, & Komppula, 2010). Today, a coastal destination is much more a complex entity (Jennings, 2004) embracing 'the full range of tourism, leisure, and recreationally oriented activities that take place in the coastal zone and the offshore coastal waters' (Hall, 2001, p.602). Even than the tourism sector's fundamental role at waterside areas is widely acknowledged (e.g. Moscardo, Pearce, Green, & O'Leary, 2001; Jennings, 2004; Cooper, 2006), these destinations generally belong to more administrative areas like countries, counties or regions (e.g. Bodensee, Lago di Garda, Lake Balaton) that influences their branding potential.

For mapping the brand equity and its elements, the case of an internationally known, 'second generation' (Chapman & Speake, 2011) waterside destination (Lake Balaton, Hungary) were chosen. The destination in focus shows the specific attributes of the Central and Eastern European context. At the first stage, the consumer-based brand equity is identified. This is followed by the media content review, and finally by analysing stakeholders' perceptions about brand equity. The results are concluded in the Five-Stage Brand Pyramid model (UNWTO & ETC, 2009).

Our main goal is to explore brand equity of a destination type (waterside) by integrating a wide range of stakeholders. The Lake Balaton's experience can provide valuable implications for similar areas (namely waterside destinations), or for destinations with an important share of domestic tourism. Furthermore, the conclusions can help in developing complex methodologies of destination brand researches.

2. TOURISM DESTINATION BRANDING

It is widely acknowledged that destination brand is a powerful tool for differentiation, and thereby providing a competitive advantage (Tasci et al., 2007; Boo et al., 2009; UNWTO & ETC, 2009). A successful brand contributes to the place identity (UNWTO & ETC, 2009), diminishes the negative effects of crisis (UNWTO & ETC, 2009; Moilanen & Rainisto, 2009), transmits value (Moilanen & Rainisto, 2009), and has a positive impact on loyalty/return visit (Usakli & Baloglu, 2011).

Today, there is trend for branding original places, nations, cultural branding, destination and place/city branding (e.g. Lorenzini, Calzati, & Giudici, 2011). At the same time, most of the branding literature is dedicated to administrative areas like countries or cities, which raises the challenge of tourism destination branding in the case of areas belonging to more administrative territories.

Acknowledging the very wide range of brand(ing) definitions, we take a look at the concept that guided us during the research process. Branding can be regarded as

- the promise of something' (Moilanen & Rainisto, 2009, p.6);
- is a competitive identity ... makes a product or destination distinctive' (UNWTO & ETC, 2009, p.6);
- destination brand is the sum of perceptions that someone has about a place which influence his/her attitudes towards that destination at an emotional level' (UNWTO & ETC, 2009, p.9). A destination brand makes the place different, exists in the eyes of the beholder (UNWTO & ETC, 2009). In this sense, the brand goes beyond to just being a logo or a slogen (UNWTO & ETC, 2009).

For the purpose of this study we defined brand (brand equity) as 'any unique attribute that makes the destination different and endows it with a special and valued character from other competitor destinations in the mind of the prospective travellers'.

The very comprehensive and colourful tourism branding literature suggests to reviewed four dimensions, namely awareness, image, quality and loyalty (Konecnik & Gartner, 2007), that is extended by value and experience (Boo et al., 2009). Tourism researchers agree that image has the main role (Lee, Cai, & O'Leary, 2006; Konecnik & Gartner, 2007) in destination branding. Brand image has the origin in destination image (Gartner, 2014), although there are a lot of ideas to measure image (originating from other sciences like e.g. pshychology), tourism marketers often differentiate between natural and man-made attributes.

Value is in the main focus of branding (Saraniemi, 2010). Besides the rational attributes, the emotional level of the brand is widely accepted (Laitamäki & Hämäläinen, 2008; UNWTO & ETC, 2009; Moilanen & Rainisto, 2009; Saraniemi, 2010; Usakli & Baloglu, 2011), extended by personality characteristics (Usakli & Baloglu, 2011). The brand has furthermore a strong connection with the travellers' lifestyle, ideal self-congruence (Nam, Ekinci, & Whyatt, 2011; Usakli & Baloglu, 2011; Barnes, Mattsson, & Sørensen, 2014).

Although acknowledging its complexity (García, Gómez, & Molina, 2012), a lot of the brand studies focus on a narrow aspect of branding (Pike, 2002). Reflecting the 'power of the travellers', consumer-based analysis (Konecnik & Gartner, 2007; Tasci et al., 2007; Boo et al., 2009) are mostly used. Recently, stakeholders' dimension (Marzano & Scott, 2009; dos Santos, 2014) came into focus, including host communities (dos Santos, 2014) or the analysis of promotional brochures (Brito & Pratas, 2015). The stakeholder perspective of branding considers the actors as decisionmakers, accepting that the destination brand is the 'outcome of the achievement of unity and collaboration among the stakeholders' (Marzano & Scott, 2009, p.249). At the same time, there are valuable case studies focusing on a mixture of stakeholders (García et al., 2012) or a special segment like attendees of conferences (Lee & Back, 2010). The complex analysis of different targets can result interesting conclusions as there are some gaps identified in the reviewed studies, e.g. perception gaps between entrepeneurs and visitors (García et al., 2012), or even between segments of travellers/heterogenity of visitors (Sahin & Baloglu, 2011). Although the intermediary stakeholders (marketing, media) have fundamental role in tourism, the branding literature gives less attention to them (e.g. among the reviewed studies by Pike (2002, 2009) there were no media targeted). This will probably change in the future, because new research methodologies become popular (e.g. content analysis).

Destination branding research is continuously developing that leads to much more focused study areas, reflecting the unique characters of each destination all over the world. Besides mapping brand equity, we can find a lot of innovative theories and frameworks like destination brand loyalty (Ekinci, Sirakaya-Turk, & Preciado, 2013), destination brand experience (Barnes et al., 2014), economic relationships (Ohe & Kurihara, 2013) or brand architecture (Datzira-Masip & Poluzzi, 2014).

Because every destination / place is different, it results a general limitation of the branding studies. Acknowledging this shortage, our main intention is to provide a valuable piece to the colourful mosaic of tourism destination branding.

3. RESEARCH METHODOLOGY

Reflecting the need for a holistic approach (Usakli & Baloglu, 2011), the study combines consumer, media and stakeholder analysis. With reference to Boo et al. (2009), the study follows the main steps of brand measurement: (1) the potential brand values are identified by customer-based brand equity, (2) the destination is compared with other competing destinations, furthermore (3) the destination in focus is well-known and popular, and a significant share of consumers interviewed had been previously to the destination. The available primary data has been analysed exclusively for this research paper.

Primary research was conducted between 2009 and 2012. The consumer survey focused on the Hungarian waterside destinations in general, the media content analysis' and stakeholder phase's target had been narrowed to the Lake Balaton. The qualitative phase following the quantitative survey provided more in-depth understanding, e.g. tourism products or activities arising form the travellers' survey were focused on at the media analysis and at the stakeholder interviews.

Data of 668 prospective domestic travellers was collected 2009, the aim was to reach the widest audience possible. The convenience sample follows the aggregated demographics of Hungarian population regarding gender and age, the tourism survey of the Hungarian Tourism Ltd. was used as a reference. The results highlight the attitudes of 'urban residents with waterside holiday experience' (see Table 1). With the aim of mapping brand equity / brand values and learning original perceptions, the following open-ended question (e.g. Pike, 2009; Sahin & Baloglu, 2011; Usakli & Baloglu, 2011) was used:

Question: According to your opinion, what are the unique characteristics of Hungarian waterside destinations, compared with international areas that you perceive as competitors?

Media content analysis is based on the Hungarian (in Hungarian language, and published in Hungary) written non-paid news. Articles were collected from the daily media observation of the Hungarian Tourism Ltd. (national tourism office). Altogether the content of 1,483 articles (published during 2009 and 2010) was monitored.

Stakeholder survey had two phases. An online questionnaire was filled out by 47 professionals (12 destination management organizations and 35 tourism service providers). The semi-structured in-depth interviews included 20 important stakeholders of the Lake Balaton's tourism (both public and private). (See Table 2)

Table 1. Consumer sample size and structure

Attribute	Sample	Hungarian population*
Gender	• Male: 47.4% • Female: 52.6%	• Male: 47.0% • Female: 53.0%
Age	• 18-24 years: 12.0% • 25-30 years: 11.3% • 31-40 years: 19.5% • 41-50 years: 19.5% • 51-60 years: 18.0% • 60+ years: 19.2%	• 18-24 years: 12.5% • 25-30 years: 13.6% • 31-40 years: 16.6% • 41-50 years: 17.0% • 51-60 years: 17.2% • 60+ years: 23.2%
Place of residence (statistical region)	• Central Hungary: 60.7% • Central Transdanubia: 6.8% • Western Transdanubia: 5.7% • Southern Transdanubia: 5.0% • Northern Hungary: 10.5% • Northern Great Plain: 2.3% • Southern Great Plain: 8.9%	• Central Hungary: 27.8% • Central Transdanubia: 11.1% • Western Transdanubia: 9.9% • Southern Transdanubia: 9.7% • Northern Hungary: 12.5% • Northern Great Plain: 15.8% • Southern Great Plain: 13.2%

Table 2. Mapping brand equity in the methodology

Consumer survey (668 persons)	Media content analysis (1,483 articles)	Stakeholders (tourism professionals) survey (47 + 20 persons)
Spontaneous associations/open-ended question	Articles' main topic Wording of the news	Spontaneous associations/open-ended question (47 persons) Semi-structured interviews valuing unique attributes (20 persons)

4. RESULTS

4.1. Lake Balaton – The Destination in Focus

Hungary can be regarded as a mature 'second generation' destination. Besides the capital city of Budapest and the health tourism resources, the Lake Balaton attracts a significant number of international visitors: 7.4% of the foreign overnight visitors have been to the lake (Hungarian Central Statistical Office [HCSO], 2013). 25.0% of the domestic overnight trips are to the Lake Balaton (HCSO, 2013).

The Lake Balaton is the greatest freshwater lake in Central Europe (Figure 1), it has a surface of 600 km2 , and its deepest point is about 11m. Being an iconic destination of the country, the region has good infrastructural location, and pleasant climate, the bath season is from May to September. The lake and its surroundings still has the strong position as a summer holiday destination (Rátz, Michalkó, & Kovács, 2008), although during the last decades, the development of 'waterindependent' attractions has come into focus (e.g. active tourism, sports, wine and gastronomy, health tourism, culture, festivals). Belonging to more administrative areas, the Lake Balaton forms one tourist region (from the altogether nine tourist regions in Hungary), its area is 3,886 km2 (4.2% of the country's total area), the volume of the population is 262 thousands (2.6% of the country's population) (HCSO, 2011). Besides the local communities and the tourists, there are a great number of second home owners in the destination (one fourth of Hungarian's second homes, namely 64 thousands units).

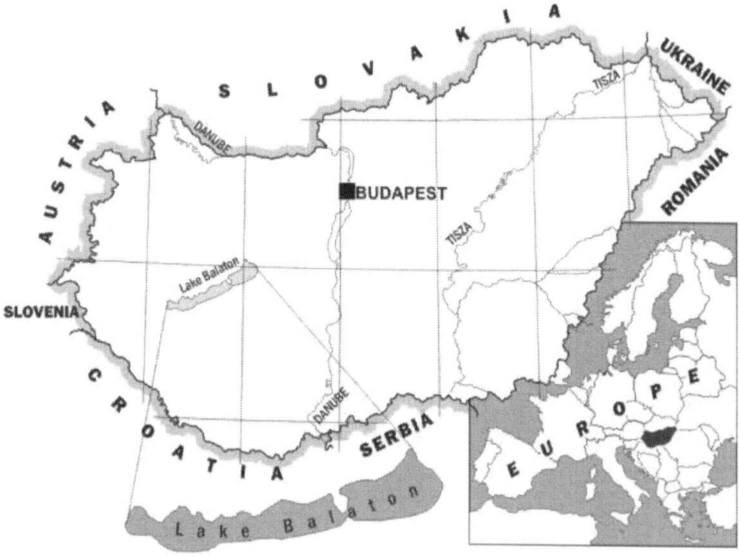

Figure 1. Geographical location of the Lake Balaton (Hungary)

The regions natural environment is protected; the total area of the Balaton Uplands National Park (founded in 1997) is 57 hectares. The region has a wide range of tourism offer: more than 50 beaches, a bike route of 200 km2 , active sport facilities, sailing facilities, health spas, museums, cultural programmes, festivals etc. are in the focus of fulfilling travellers' motivations. The destination is mainly managed by local tourism organizations (destination management organizations, DMOs); the regional destination management organization was founded in 2011.

The destination's commercial accommodation establishments register app. 5 million guest nights (4.8 million guest nights in 2013, of that 62% domestic guest nights) yearly (HCSO, 2013). Tourism demand to the destination was characterized by restructuring at the beginning of the 1990's. Tourism at the Lake Balaton is still characterized by a strong seasonality, a high concentration to the coastal area, and the dominance of the domestic overnights. Furthermore, it should be highlighted that the area is a very popular summer holiday destination, and an iconic tourism receiving place of the former socialist countries in the Central and Eastern European region.

4.2. Customer-Based Brand Equity

621 respondents out of the 668 interviewed persons shared their views about the brand equity. In the first step of the analysis, the mentioned attributes were grouped into 23 categories. Most of the mentioned assets are positive, however there are negative attitudes (lower quality compared to foreign destinations, lack of hygiene, not exciting enough or crowded) that could be a source of conflict, and thereby resulting unsatisfied visitors (see Table 3).

The brand image elements can be split into four main dimensions: besides the destination type's (namely waterside) assets, the benefits of domestic travel, the tourism products and activities, furthermore the emotions and other items were identified. The 'landscape' itself (natural values including flora & fauna, climate, mosquitos, furthermore the freshwater's advantages against sea water) is very attractive for tourists escaping from the daily routine. The second dimension has a strong connection with the consumers' country of origin that means the perceived advantages (closeness, safety, and cheap prices) of domestic travel in this case. The next dimension covers the colourful experiences, the wide range of activities available in the destination, like cultural tourism, health tourism, gastronomy and wines, fishing, and active tourism. Finally, the emotional dimension (former visits, experiences, ambience, nostalgia) also stay in the foreground of the prospective travellers' perceptions. (See Figure 2)

Table 3. Consumer-based brand equity elements

Element	Number of responses
Natural treasures (landscape, flora and fauna, climate)	122
Close, accessible	113
Unique places, memories, ambient, nostalgy	79
Safe / feeling home (e.g. language), get to know national treasures	78
Cheap(er than a lot of foreign destinations)	66
Freshwater (e.g. swallow, appropriate for families with small children), grassy beaches	51
Cultural monuments, programmes	42
Quiet, good for relaxation	42
Health tourism	41
Low level of service quality	35
Hospitality	31
Gastronomy	30
Dirty, disordered	29
Expensive (than a lot of foreign destinations)	16
Fishing	15
Wide range of facilities, 'fit for all'	15
Not unique	14
Not exciting enough	11
Active tourism, sports	9
Lake and sea cannot be compared	9
Wine tourism	7
Crowded, a lot of tourists	7
Mosquitos	2

With reference to the brand quality, we can conclude that the potential travellers are very priceconscious, the destination in focus is perceived rather cheap than expensive (compared with international tourism receiving areas in the mind of potential visitors).

4.3. Brand Values Interpreted by the Media

The potential brand values interpreted by the media had been studied by monitoring the main topic and the article's wording, and by mapping the seasonal variance of main topics covered in the news. In the media interpretation, the destination is attractive for a wide range of segments; the media acknowledges the colourful travel experience, the wide range of attractions and activities. The colourful tourism offer means a great challenge for its positioning and branding. Probably in order to call readers' attention, the media often uses 'the most/biggest/longest/oldest...' terms.

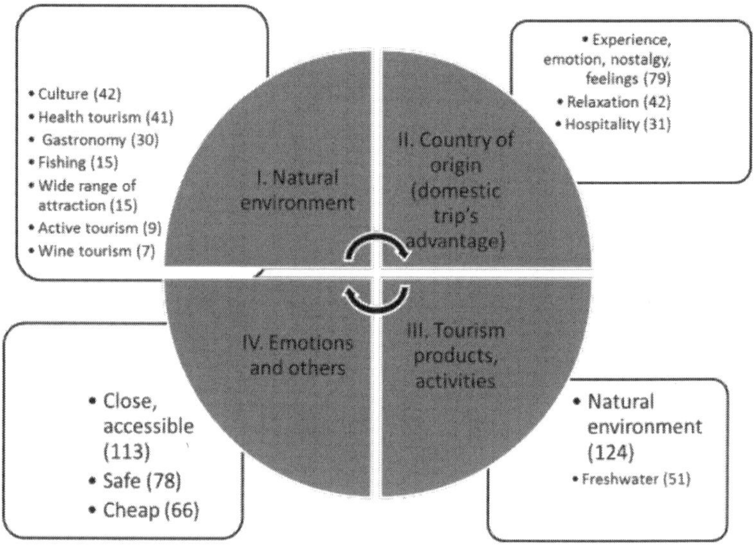

Figure 2. Hungarian waterside destinations' (Lake Balaton's) consumer-based brand equity (number of responses)

The articles' wording shows a positive image of the destination, highlighting its landscape/scenic beauty. The destination's natural environment (91 of the altogether 800 topics) can be found throughout the year, and independently from travellers' motivation/segments (e.g. articles about active tourism also highlight the natural treasures). The domestic focus includes the terms 'Hungarian sea', 'national treasure'. Here we should also highlight the emotional dimension of brand equity that means mainly the nostalgia in the media news. From the wide range of tourism products and activities (based on the number of topics covered in the news), cultural tourism has the widest supply (115 topics). It is followed by active tourism (64 topics), wine & gastronomy (49 topics), finally health tourism (19 topics).

Meanwhile cultural tourism has something to offer throughout the year, active tourism is the most attractive from Spring to Autumn (following the traditional travel season), and health tourism news are in the focus from Autumn to Spring (during the Summer, there were no health tourism related article among the studied news). The media news about tourism products show a seasonal distribution that highlights the limitations of these products, and the necessity of harmonizing with the destination assets, in order to be developed to a real competitive advantage/brand value. (Figure 3)

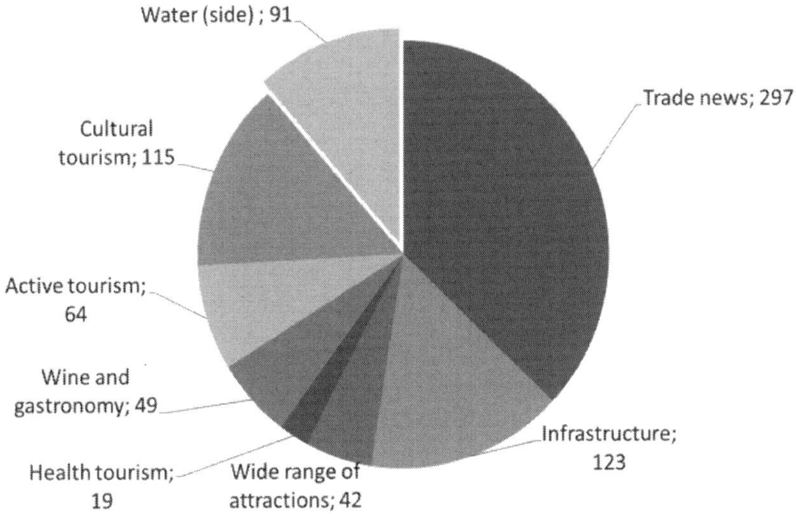

Figure 3. Main topics of the media news

4.4. Stakeholder-Based Brand Equity

Stakeholder-based brand equity has been surveyed among tourism professionals. Because of the destination's international reputation, we differentiated among brand values on domestic and on international level. From the four dimensions identified at the consumer survey, the destination specific attributes (landscape, freshwater, the lake's shallowness, the big surface, and the water's healing effect) play a fundamental role. The domestic trip's advantages (e.g. accessibility) are also at the core of brand values among prospective domestic travellers. Among the wide range of tourism products, active tourism is the most valued by stakeholders. Wine tourism, family-friendly services and fishing are of greater importance for domestic guests. Health tourism is valued only partly as a brand value, as it attracts a smaller/specific segment. So sailing and cycling are also less attractive for a wide range of segments. The one and only emotional value highlighted by the tourism professionals is the nostalgia. On the domestic level, there is a focus on the wide range of attractions, meanwhile on the international level, the gastronomy and the good value for money can be a more important brand equity item. (Figure 4)

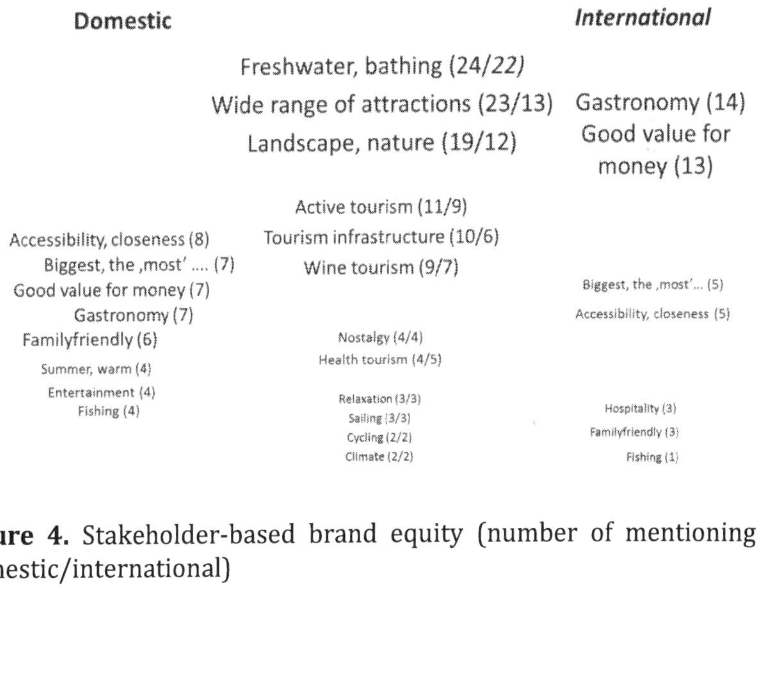

Figure 4. Stakeholder-based brand equity (number of mentioning - domestic/international)

4.5. Summary of Research Results

With the aim of integrating brand equity among different target groups (consumer, media, tourism stakeholders), we applies the Five-Stage Brand Pyramid model (UNWTO & ETC, 2009). The results demonstrate the strong dominance of landscape, including 'lacustrine' specific asset, like the freshwater's advantage against sea water (destination specific attributes). The natural environment can be found in all of the stages of the pyramid. This is supported by the country of origin dimension's perceptions, namely the perceived advantages of domestic travel.

At the bottom of the pyramid, among the rational attributes we find good quality beaches, accessibility, nice weather, and all of the non-water-based tourism products. It points out that tourism products (active tourism, cultural tourism, health tourism, wine and gastronomy) should not be dedicated outstanding role as a brand value, because these activities can motivate smaller segments. Among the emotional benefits (the pyramid's next level) there are the benefits of domestic trip, the personal experiences, furthermore the time spent with family & friends. The Lake Balaton's brand personality is built on bathing in freshwater, quiet environment, a wide range of activities, and freedom. These attributes can differentiate the lake from its perceived – domestic or foreign – competitors. On the fourth level of the pyramid, we find a positioning statement (not a slogan!) representing the most dominant attributes. Besides the benefits of domestic travel, we should here highlight that the Lake Balaton and its surroundings is not a 'once-in-a-life' destination, a lot of the visitors return to the area from time to time. Among the potential brand values, no iconic attraction (settlement or sight) emerged.

The destination's core assets are the human-scale landscape, domestic environment (safety), and a wide range of colourful experiences. (Figure 5)

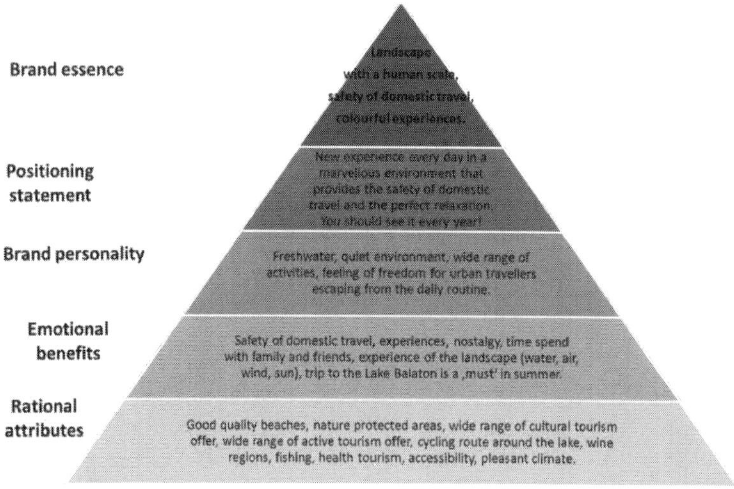

Figure 5. The Lake Balaton's potential brand values in the five-stage brand pyramid model

Compering perceptions of different target groups, our research highlighted some important gaps to be considered. Generally, travellers' put much more emphasis on the emotional dimensions that is only partly supported by the professionals/supply side (nostalgia was the common issue).

Table 4. Destination brand equity dimensions among different stakeholders

Brand equity element	Potential travellers/Demand side (n=668)	Media (n=1,483)	Tourism professionals/Supply side (n=47+20)
Destination specific attributes (natural environment)	Natural environment, landscape, climate, freshwater	Scenic beauty	Human-scale landscape Water itself (freshwater, shallowness, healing effect)
Demand specific attributes (domestic travel, attitudes towards destination, towards tourism)	Close Safe Cheap	'Hungarian sea', 'national treasure'	Accessible, close
Tourism products, activities	Cultural tourism Wine & gastronomy Health tourism Active tourism	Cultural tourism Active tourism Wine & gastronomy Health tourism	Active tourism Wine tourism Health tourism
Emotions	Former experience, emotions, nostalgia	Nostalgia	Nostalgia
Others	Wide range of attractions	'The most/biggest/longest/oldest…' Wide range of attractions	Wide range of attractions Good value for money Family friendly

The destination specific attributes (originating from the destination type like waterside, mountain, rural or city) is strongly acknowledged by target groups. The country of origin specific items (domestic travel) is dominated by accessibility, followed by safety, and affordable price level. The tourism products' potential as a brand value is not perceive equally by the stakeholders. Meanwhile cultural tourism appears strongly in the mind of visitors and in the media, the supply side does not consider it as a brand equity item. On the other side, visitors do not think active tourism to be a unique, differentiating brand value of the destination. (Table 4)

5. THEORETICAL AND PRACTICAL IMPLICATIONS

Our research resulted in some valuable implications about tourism destination branding for tourism stakeholders, destination managers. The consumer-based brand equity elements could be grouped into four main dimensions: (1) destination specific assets, (2) country of origin values (domestic travel), (3) activities, and (4) emotional dimensions and other attributes. The suggested brand equity (image) elements originating from the consumer survey were partly supported by the media and the stakeholder survey. With reference to other researches mapping the role of source markets, our study acknowledged the rational and emotional benefits of domestic travel (that can be of relevance for almost all the destinations, as most of them receive a high share of domestic visitors).

From the practical point of view, the study has the following implications:

- Using a complex methodology / holistic view, that embraces a wide range of target groups enables to identify potential gaps between the perceptions of demand and supply side.
- Consumers are much more emotion-driven that should be considered during the process of destination branding.
- Tourism products' / activities' potential as a core asset need a comprehensive research in order to result real brand values, and not 'copy-cat' destinations.

6. LIMITATIONS OF THE STUDY

The researched area, the Lake Balaton (Hungary) is a mature, second generation lacustrine destination. This has the advantage of the potential analysis of a wide range of actual issues. The conclusions' implication for other areas need further thinking and has to take into account the specific characteristics of each destination.

The sample was homogenous in terms of demographic – waterside experience, urban travellers. So, any conclusion regarding the heterogeneity of travellers should be viewed under this limitation. Our study included the domestic dimension. For destinations with a high share of international tourism demand, the conclusions' implication should be extended with the international dimension.

ACKNOWLEDGEMENTS

Consumer survey was founded the Hungarian Scientific Research Fund (OTKA, project number K100953). Media content analysis, online survey among tourism professionals, and in-depth interviews with stakeholders were founded by the Authors

REFERENCES

1. Agarwal, S. (1997). The resort cycle and seaside tourism: an assessment of its applicability and validity. Tourism Management, 18(2), 65-73.

2. Agarwal, S. (1999). Restructuring and local economic development: Implications for seaside resort regeneration in Southwest Britain. Tourism Management, 20(4), 511-522.

3. Barnes, S. J., Mattsson, J., & Sørensen, F. (2014). Destination brand experience and visitor behaviour: Testing a scale in the tourism context. Annals of Tourism Research, 48, 121-139.

4. Boo, S., Busser, J., & Baloglu, S. (2009). A model of customer-based brand equity and its application to multiple destinations. Tourism Management, 30(2), 219-231.

5. Brito, P. Q., & Pratas, J. (2015). Tourism brochures: Linking message strategies, tactics and brand destination attributes. Tourism Management, 48, 123-138.

6. Buhalis, D. (2000). Marketing the competitive destination of the future. Tourism Management, 21 (1), 97-116.

7. Chapman, A., & Speake, J. (2011). Regeneration in a mass-tourism resort: The changing fortunes of Bugibba, Malta. Tourism Management, 32(3), 482-491.

8. Cooper, C. (2006). Lakes as tourism destination resources. In C. M. Hall & T. Härkönen (Eds.), Lake tourism: An integrated approach to lacustrine tourism systems (pp. 27-42). Clevedon: Channel View Publications.

9. Datzira-Masip, J., & Poluzzi, A. (2014). Brand architecture management: The case of four tourist destinations in Catalonia. Journal of Destination Marketing and Management, 3(1), 48- 58.

10. dos Santos, E. R. M. (2014). Destination mutant brands: An empirical perspective of tourism professionals and host community to create Mutant Brand Cape Verde. Journal of Destination Marketing and Management, 3(1), 68-78.

11. Ekinci, Y., Sirakaya-Turk, E., & Preciado, S. (2013). Symbolic consumption of tourism destination brands. Journal of Business Research, 66(6), 711-718.

12. García, J. A., Gómez, M., & Molina, A. (2012). A destination-branding model: An empirical analysis based on stakeholders. Tourism Management, 33(3), 646-661.

13. Gartner, W. C. (2014). Brand equity in a tourism destination. Place Branding and Public Diplomacy, 10, 108-116.

14. Hall, C. M. (2001). Trends in ocean and coastal tourism: the end of the last frontier? Ocean & Coastal Management, 44(9-10), 601-618.

15. Hungarian Central Statistical Office [HCSO] (2011). Jelentés a turizmus 2011. évi teljesítményéről. Hungary: Budapest.

16. Hungarian Central Statistical Office [HCSO] (2013). Jelentés a turizmus 2013. évi teljesítményéről. Hungary: Budapest.

17. Jennings, S. (2004). Coastal tourism and shoreline management. Annals of Tourism Research, 31(4), 899-922.

18. Konecnik, M., & Gartner, W. C. (2007). Customer-based brand equity for a destination. Annals of Tourism Research, 34(2), 400-421.

19. Konu, H., Tuohino, A., & Komppula, R. (2010). Lake wellness - A practical example of a new service development (NSD) concept in tourism industries. Journal of Vacation Marketing, 16(2), 125-139.

20. Laitamäki, J. M., & Hämäläinen, T. J. (2008). Drivers of globalization in the tourism industry: Global strategies for country destination brands. In K. Lindroth & M. Voutilainen (Eds.), Competition in tourism: Business and destination perspectives. Paper presented at the Proceedings of the Travel and Tourism Research Association Europe 2008 Annual Conference in Helsinki, Finland, 23-25th April (pp. 83-96). Helsinki: HAAGA-HELIA University of Applied Sciences, on behalf of TTRA Europe.

21. Lee, G., Cai, L. A., & O'Leary, J. T. (2006). www.branding.states.US: An analysis of brandbuilding elements in the US state tourism websites. Tourism Management, 27(5), 815-828.

22. Lee, J. S., & Back, K. J. (2010). Reexamination of attendee-based brand equity. Tourism Management, 31(3), 395-401.

23. Lorenzini, E., Calzati, V., & Giudici, P. (2011). Territorial brands for tourism development: A statistical analysis of the Marche Region. Annals of Tourism Research, 38(2), 540-560.

24. Marzano, G., & Scott, N. (2009). Power in destination branding. Annals of Tourism Research, 36(2), 247-267.

25. Moilanen, T., & Rainisto, S. (2009). How to brand nations, cities and destinations - A planning book for place branding. Great Britain: Palgrave Macmillan.

26. Moscardo, G., Pearce, P., Green, D., & O'Leary, J. T. (2001). Understanding coastal and marine tourism demand from three European markets: Implications for the future of ecotourism. Journal of Sustainable Tourism, 9(3), 212-227.

27. Nam, J., Ekinci, Y., & Whyatt, G. (2011). Brand equity, brand loyalty and consumer satisfaction. Annals of Tourism Research, 38(3), 1009-1030. [28] Ohe, Y., & Kurihara, S. (2013). Evaluating the complementary relationship between local brand farm products and rural tourism: Evidence from Japan. Tourism Management, 35, 278- 283.

28. Rátz, T., Michalkó, G., & Kovács, B. (2008). The influence of Lake Balaton's tourist milieu on visitors' quality of life. Tourism: An International Interdisciplinary Journal, 56(2), 127- 142.

29. Pike, S. (2009). Destination brand positions of a competitive set of near-home destinations. Tourism Management, 30(6), 857-866.

30. Pike, S. (2002). Destination image analysis – A review of 142 papers from 1973 to 2000. Tourism Management, 23(5), 541-549.

31. Sahin, S., & Baloglu, S. (2011). Brand personality and destination image of Istanbul. Anatolia: An International Journal of Tourism and Hospitality Research, 22(1), 69-88.

32. Saraniemi, S. (2010). Destination brand identity development and value system. Tourism Review, 65(2), 52-60. [34] Tasci, A. D. A., Gartner, W. C., & Cavusgil, S. T. (2007). Measurement of destination brand bias using a quasi-experimental design. Tourism Management, 28(6), 1529-1540.

33. UNWTO. (2011). UNWTO tourism highlights (2011 ed.). Madrid: World Tourism Organization. Retrieved from http://mkt.unwto.org/sites/all/files/docpdf/unwtohighlights11en hr.pdf. [36] UNWTO., & ETC. (2009). Handbook on tourism destination branding. Madrid, Spain: World Tourism Organisation.

34. Usakli, A., & Baloglu, S. (2011). Brand personality of tourist destinations: An application of self-congruity theory. Tourism Management, 32(1), 114-127.

35. Yeoman, I. (2007). Tomorrow's tourist: Scenarios & trends. Elsevier.

3

THE GWALIOR FORT: CHALLENGES AND OPPORTUNITIES

Shikha Sharma

Institute of Vocational Studies, Masters of Tourism Administration, Himachal Pradesh University Shimla, India

ABSTRACT

Heritage Tourism is one of the fastest and prominent sectors in comparison to the other form of tourisms. Heritage tourism focuses on destinations, natural and cultural values and goes beyond a simple interest in the past. Heritage and cultural tourism encompasses a wide variety of landscapes and settings, it explores the cultural and natural heritages of peoples, highlighting natural physical beauty, urban or industrial developments. Heritage of destination like Gwalior is rich in heritage and had made contribution towards its past. The Gwalior fort is facing many challenges at present but at the same time the scope for opportunities is no less. A proper model should be followed for the protection, preservation and conservation of this national monument as this can be a major attraction which can help in placing Gwalior amongst the most favourable destinations .This paper tries to highlights different aspects related to monument, traditional scenery, handicraft, food, accessibility, preservation, museum, architecture, light and sound show, live shows which can help in the development of the fort.

KEYWORDS

Heritage, Gwalior fort, challenges, opportunities.

1. INTRODUCTION

The three words i.e. culture, heritage and tourism are extremely correlated to each other. However culture, heritage and tourism have their own independent identity and existence but at the same time each is dependent on the other as well. Culture cannot exist without heritage and where there is culture and heritage, tourism will have its existence. During the past decade cultural and heritage tourism has undergone tremendous growth and cultural and heritage tourism according to WTO (Richards, 2000) is a very important segment of the tourism demand which represents 37% of international tourism. Boniface and Fowler call tourism the greatest show on the earth (1993, p. 11) whose greatest ingredient is heritage.

The glorious past of India has also ensured that the present and subsequent generations have plenty of historical and cultural heritages to be proud of. For centuries, people migrated into rich country. India with her respective culture and tradition later mixed with the rituals of the Indians has given birth to a new colour of heritage and culture. This amalgamation of cultures has slowly but gradually has enriched the values and faith of the Indians even further with each sunrise and sunset in the sky. The rich heritage of India comes alive with all its fragrance and colour which are still breathing in the heritage palaces, forts, temples and monuments fringed in each and every corner which makes multifaceted country, India. The splendid architecture, minutely carved motifs, embellished facades of the heritage monuments narrates the immense skill of the craftsmen of the country which have been fascinating the world for many centuries. If one start counting may fail to accommodate all the heritage sites of India in fingers which have evolved with the passage of time. UNESCO has come up with the mission to protect these treasures which are finest gifts to world from India so that generations to come feel them as fortunate. The number of important monuments and sites in India is so great that a single tour to discover the multifarious facets of its wonderful heritage is not enough. Nevertheless it will give an admirable glimpse into the Indian History and Heritage. Set with this background this research examines the satisfaction of tourists visiting the heritage sites of Gwalior.

2. LITERATURE REVIEW

Tourism is Basic and most desirable human activity deserving the praise and encouragement of all peoples and all governments. It is an industry concerned with attracting people to a destination, transporting them, housing and entertaining them upon arrival and returning them to their homes. Travel, today is sold like a normal consumer product through retail outlets, wholesalers and even departmental stores of many countries. More and more people with small incomes are taking overseas travel possible. In addition to leisure time activity, tourism has many faces like cultural tourism, religious tourism, eco-tourism, heritage tourism, adventure tourism, rural tourism, leisure tourism, beach tourism, social tourism, medical tourism etc. A lot of emphasis has been laid down on the economic role of tourism, its contribution to foreign exchange earnings of the nation its potential for greater employment. It is the time to understand the important role of tourism as a means of social education and social mobilization in the country itself.

Heritage is a legacy for our past, a living and an integral part of our life today. Heritage is defined as cultural and natural heritage. Cultural heritage can be described in terms of both tangible and intangible. Heritage can be termed as living interpretations of the past in present.

Tangible Heritage encompasses the vast created works of humankind including places of human habitation, villages, towns and cities, buildings, structure, art works, documents, handicrafts, musical instruments, furniture, clothing and items of personal decoration, religious, ritual and funerary objects, tools, machinery and equipment and industrial systems.

Tangible Heritage encompasses -Movable heritage (like paintings, sculptures, coins, manuscripts etc.), Immovable Heritage (monuments, archaeological sites and so on) and Natural Heritage.

Visits to historic sites, including those that portray events associated with human suffering and mass death, have become a significant aspect of tourist visitation (Waitt, 2000), for reasons such as opportunities for visitors to re-live those historical events and for the monetary value of such visits for conservation purposes. Examples include the slave forts and castles in the coast and war related sites such as Museums with exhibits on the Holocaust and Cuchi Tunnels of Vietnam (Henderson's, 2000). Foley and Lennon (1996) and Seaton (1996), focusing on sites associated with the death of important contemporary history, have described the phenomena as being visited by people who are fascinated by death. Rojek (1993) have described the visit to these sites as private enjoyment of pleasure in the events associated with those sites, at least for some visitors. However, although it is possible that some visitors may be fascinated by the 'gruesome', such sites are rarely developed to serve the pleasure needs of visitors, but some entertaining activities may be included to heap visitors' numbers. We therefore find that in some (Garrod & Fyall, 2000).However, the presentation and interpretation of heritage sites are susceptible to distortions (Boniface and Fowler, 1993) and they are powerful sources of misunderstanding and liberations of strong emotions (Fowler,1992).

Graham et.al. (2000) describe Heritage as cultural and economic capital, vulnerable to exploitation of various types. Governments responsible may be following hegemonic aims (Tunbridge 1984; Mc Intosh and Prentice (1999), for e.g. describe the search for meaning and pride in the past as the result of the divorce of western societies from their rural origins through urbanisation and population drift to the cities. Institutions such as Heritage attractions and museums play a key role in conferring authority upon their interpretation of the past. Following this general line of argument Prideaux and Kininmount (1999; p-300) observed that one of the roles of Australian heritage attractions, including rural museums, was to act as 'a repository of many of the artefacts that embody Australia's image of itself as land of battles, mates an mate ships, a fair go for all and victory in the fall of adversity. In many non -Beach holiday destinations', heritage attractions may constitute important tourism resources, one e.g. being the tourist-historic city (Ashworth & Tunbridge, 1990; Timonthy &Wall 1995; Prentice et.al. 1998).Heritage attractions in an area differs from other attractions such as theme parks, shopping precincts /national areas. Richards (2000,p-9),for e.g. states that Heritage tourism is largely concerned with the cultural legacy of the past , or the 'hard' cultural resources usually contained in old buildings, museums, monuments, and landscapes or represented and interpreted in specialised Heritage centres. In a broadly similar manner, Ashworth (2000p, 19) defined Heritage tourism as the 'co modified artefacts, buildings, memories and experiences of the past that entails cooperation between heritage producers, the tourism industry and the local place managers. Light et al (1994) describes Heritage tourism as belonging to the middle -class, well -educated, middle -aged, no children, on holiday away from home and who have a prior knowledge of history. This view is largely supported by Balcar & Pearce (1996) and Prentice (1993). Identification of target markets based on preference patterns of tourist for specific types of Heritage attractions is essential as not all heritage tourists can be expected to be interested in all Heritage attractions. Previous research have revealed that Heritage tourists are likely to spend more money per trip than other groups of tourists (Kerstetter et. al; 2001) although the time spent at heritage sites is disputed, with Kerstetter et.al (2001) quoting the average length of stay as days whereas Ashworth (2000) argues for a shorter length of stay.

2.1 Cultural/Heritage Destination Attributes

The aim of the study is to identify the attributes of the Gwalior fort which satisfy the tourists when they visit it. According to Andersen, Prentice and Guerin (1997), the several attributes are historical buildings, museums, galleries, theatres, festivals and events, shopping, food, palaces, famous people, castles, sports, and old towns in a destination. Richards (1996) chose several attributes related to cultural/heritage destinations in order to analyze European cultural tourism. Glasson (1994) explained the impacts of cultural/heritage tourism and management responses through an overview of the characteristics of tourists to Oxford. Peleggi (1996) examined the relevance of Thailand's heritage attractions to both international and domestic tourism, including an analysis of the state tourism agency's promotion of heritage and the ideological implications of heritage sightseeing in relation to the official historical narrative. This research provided several attributes, such as traditional villages, monuments, museums, and temples.

Tourist satisfaction is important to successful destination marketing because it influences the choice of destination, the consumption of products and services, and the decision to return (Kozak & Rimming ton, 2000). Several researchers have studied customer satisfaction and provided theories about tourism (Bram well, 1998; Bowen, 2001). For example, Parasuraman, Zeithaml, and Berry's (1985) expectation-perception gap model, Oliver's expectancy- disconfirmation theory (Pizam and Milman, 1993), Sirgy's congruity model (Sirgy, 1984 ; Chon and Olsen, 1991), and the performance - only model.(Pizam, Neumann, and Reichel, 1978) have been used to measure tourist satisfaction with specific tourism destinations.

Furthermore, after tourists have bought the travel service and products, if the evaluation of their experience of the travel product is better than their expectations, they will be satisfied with their travel experience.

2.2 Gwalior and Gwalior fort

The Gwalior district, which is in the Gwalior revenue division, lies in the northern part of the State .It covers an area of 2002 square miles which is a little more than 1.1 percent of the State. The district has a population of 1,629,881, according of the Census of 2011, distributed almost equally between the rural and urban areas. The Gwalior is bounded on the north and North West by Morena district, on the North -east by Bhind district, on the east by Datia district and on the south by the Shivpuri district. Gwalior was founded by a Kachhwaha Chief, named Suraj Sen., the Petty Raja of Kuntalpuri. Suraj Sen. was a leper, and one day when thirsty with hunting near the hill of Gopagiri, he came to the cave of the Sidh Gwalipa, and asked for water. The hermit gave him some water in his own vessel, and no sooner had he drunk it than he was cured of his leprosy. The grateful Raja then asked what he could so for the holy man, and he described to built a fort on the hill, and to enlarge the tank from which the healing water had been drawn. Suraj Sen. accordingly built the Fortress, which he named after the hermit Gwali-awar, or Gwalior, as it is now written.

Gwalior has a rich historic background. Gwalior has tangible heritage, living heritage, natural heritage, archaeological heritage, and pilgrim heritage, and intangible heritage-Tansen mahoutsav, Gwalior mela, Gwalior cuisine.

The fort of Gwalior, aptly called the author of Taj –ul – Maasir, the 'Pearl in the necklace of the castles of Hind' stands on a long and narrow rocky hill of sandstone which rises abruptly 300 ft above, the surrounding country. It is a mile and three quarters long, running north to south, and varies in width from 600 ft. to 2800 ft. east to west. In some places the cliff overhangs and in others, where the hill sides are less precipitous they are scraped. The wall of the fortress is 30 to 35 feet high with the rock below them steeply scarped down to a sloping glacis of crumbling basalt.

The fort contains various objects of historical and antiquarian interest. Today two roads climb up to the fort- one to the Gwalior Gate on the east and the other to the Urwahi Gate on the west. The first path is defended by five gates, placed at strategic points along its course. Originally there were two more gates but they have been removed. The fort possesses numerous tanks, besides two rock cut cisterns. There are various monuments in the fort, including six palaces, four Hindu and Two Muslim.

3. METHODOLOGY

The methodology adopted for the study was observation and interview method. The Gwalior fort was visited randomly 3 times and tourists were interviewed each time on the aspects like challenges faced by the fort and the opportunities for improvement. The observation method was adopted as it suits best in this situation.

3.1 Research Questions

The question was to find out the exact condition of the fort in present scenario. The concern was to find out different ways in which it could be made famous and well preserved for the future. The main concern was to find out the challenges and the opportunities there are for the development of the area in and around the Gwalior fort.

3.1.1 Gwalior Fort: The Challenges

Monument - The condition of the monument is in bad shape. The fort is under the archaeological survey of India but it has not been well preserved and conserved. Though an entry fee of Rs .5/ is taken from the Indian nationals but the monument is not under properly managed hands. This massive monument has the capacity to attract a large number of tourists but as we head toward the surroundings of the monument are in bad shape. From the main entrance gate to the other entrance gate there is a long way up the hill which has not been well preserved and managed. Clean and green environment around the monument would have added to its beauty but there is a quick entrance to almost everybody in this area. Inside the monument there are some areas which are not electrified and have dark rooms. The tourists are not able to visit these areas until and unless they hire a guide or they have some lightning with them. The dark portion of the monument which lies underground becomes unapproachable causing inconvenience to the tourist and thus increasing their irritation level and negativity for the fort. The fort is not clean at some places. The fort is huge and has a significant place in the history of Gwalior but it has no environment to stay there for long.

Traditional scenery: The fort is massive but there is no such beautiful scenery around it. It is at the highest level and the entire city can be viewed from here standing at the walls of the fort. But the traditional beauty and scenery has been lost due to unaware attitude of the locals and the tourists.

Handicrafts: There are no outlets of the handicrafts around the fort. Tourist visiting the fort has some kind of expectation as Gwalior is famous for its textile and cloth industry. Being under the ASI and having a reasonable high area under it there is no shop of handicraft which could have portrait the local art and culture to the tourist and which would have satisfied the tourist in a way that there was something for them to be carried away as souvenirs especially for those who have come to the fort with expectations high. Government have not taken any initiative to open any outlet for the promotion of local art and craft of the place. The fort would have been the best source point for a cultural activity like this.

Food: There is no proper provision for food courts or restaurants in and around fort. The number of shops providing basic facilities like mineral water, tea, coffee are just negligible. The tourists who come to a fort have high expectations and deserve some kind of facilities. If there would be such facilities in the fort to a quality wise end it will probably help the local community to take interest and participate in making the fort more beautiful and bounty. The locals are also going to have source of income and they can get some money for their household.

Accessibility: There is a problem of accessibility in terms of govt. run buses. There are no buses which could run on certain routes to take the tourists or the locals to the fort directly. Going to the fort is a big challenge if one is not interested in hiring a vehicle which could directly take a person to the fort. But at the same time, if tourists want to explore the fort on their own they will face many problems –like the conditions of the road is not good, there is no other mode to go to the fort except own vehicle, hired vehicle .if one is interested in going by its own the person has to change 2-3 or even more vehicles to reach to the fort. Thus, reaching the fort is a big challenge in terms of uneasiness caused to the tourists

Preservation: In order to pass on the heritage to the future generations what is currently identified as being culturally significant today, we must imbibe good conservation practices for fort in order to prevent it from deterioration and extend the life and basic functions of the fort. The various threats such as environmental threats –moisture, intense solar radiations, prevailing winds, rainfalls which change the physical attributes of the fort should be taken into account

3.1.2 Gwalior Fort: The Opportunities

Museum: The museum is present near the fort which provides an opportunity to see the artefacts which have relevance and are a part of the history of the state. The museum should be enlarged by keeping some articles for sale which are of tourist interest and which have a higher satisfaction level for the tourists. The Museum can sell souvenirs and handicraft related items

Architecture: The architecture of the Fort is so massive that it is clearly visible from different areas in the destination and also during the rail journey approaching towards Gwalior. One can easily feel the royal and grand structure as one starts approaching Gwalior. It is so prominent that it creates anxiety in the mind of the traveller as to which fort it could be. Its location in terms of its visibility from far off places provides an opportunity for the Government to make it more prominent in eyes of tourists by promoting it and improving its surroundings.

The light and sound show: The light and sound show already held in the vicinity of the fort should improve in the quality and variety should be offered to the tourist. The same show is running since the last many years. There has been no change in the style the show is conducted.

Live show: There could be a provision of live shows where the local community can be made a part of it. The locals can depict their art and culture in the form of dance and music. The folk songs and folk dances of Madhya Pradesh can be performed in front of the tourists and this will also boost the local's art and culture. At the same time it will also provide some kind of community participation in the development of the forts image. it will also provide satisfaction to the tourists.

4. CONCLUSION

From the above studies it has been concluded that the Gwalior fort hold an important and significant part in the history of Gwalior. It has great heritage value and it comes under the category of built heritage. Gwalior fort faces some major problems in present scenario which are not good for its image. The fort has certain shortcomings which should be covered by the state government or the Archaeological survey of India collectively. It should be viewed as an important tourist destination as it has and it is attracting large number of tourist from different parts of the country but the focus should be to attract more tourists from international market also. It should bring some amendments in the way the present scenario is going on and at the same time challenges and opportunities should be sought out. It is great heritage of the state and it should cater to the tourist need and expectations in the right way. proper steps should be taken to conserve the beautiful piece of fine architecture so that it can help to uplift the states image and be fruitful for the community as well.

REFERENCES

1. Ashworth G, J, TunbridgeJE.(1990). The Touristic Historic City .Belhaven : London.

2. Ashworth G.J (2000). Heritage ,Tourism and Places :A Review .Tourism Recreation Research ,25(1):19-29

3. BalcarM, PearceD.G.1996. Heritage Tourism on the West Coast Of New Zealand. Tourism Management.17(3):203-212.

4. Bram well, B. (1998). User satisfaction and product development in urban tourism. Tourism Management, 19(1), 35-47

5. Boniface.P.,Fowler,P.J., (1993) Heritage and Tourism : In the Global Village. London, Routledge

6. Foley M., Lennon. J.J. (1996). Heart of darkness .International Journal of Heritage Studies 2(4):195-197.

7. Fowler .P.J.(1992). The Past in Contemporary Society: Then, Now. Routledge: London.

8. Garrod B., Fyall. A. (2000).Managing Heritage Tourism .Annals of Tourism Research 27(3): 682-708

9. Glasson, J. (1994). Oxford: a heritage city under pressure. Tourism Management, 15(2), 134- 144.

10. Henderson .J.C. (2000). War as tourist attraction: the case of Vietnam. International Journal of Tourism Research .2(4):269-280

11. Hewison R. (1987). The Heritage Industry: Britain in a Climate of Decline .Methuen: London.

12. Kozark, M. & Rimmington, M. (2000). Tourist satisfaction with Mallorca, Spain, as an offseason holiday destination. Journal of Travel Research, 38, 260-269

13. Lights D, Prentice C, Ashworth .G.J. Larkham J. (1994).Who consumes the heritage products?Implications for European Tourism.In Building New Heritage: Touirsm, Culture & Identity.in the New Europe., Routledge:London.90-116.

14. Mc.Intosh A.J,Prentice R.C.(1999).Afirming authenticity consuming cultural heritage, Annals of Tourism Research 26(3):589-612.

15. Parasuraman, A., Zeithaml, V.A. & Berry, L (1985). A conceptual model of service quality and its implications for future research. Journal of Marketing, 49(Fall), 41-50.

16. Peleggi:, Maurizio (1996). National heritage and global tourism in Thailand. Annals of Tourism Research, 23(2), 340-364.

17. Pizam, A., Neumann, Y. & Reichel, A. (1978). Dimensions of tourist satisfaction with a destination. Annals of Tourism Research, 5, 314-322.

18. Pizam, A. & Millman, A. (1993). Predicting satisfaction among first-time visitors to a destination by using the expectancy-disconfirmation theory. International Journals of Hospitality Management. 12(2), 197-209.

19. Philips & Tunbridge J .E.(1994).New supports for Heritage Tourism in rural Ireland. Joural of Sustainable Tourism 2(1&2):112-129.

20. Prentice R, Guerin S, Mc Gugan S. 1998 . Visitors Leaarning at a heritage attractions:A Case study of discovery as a media product.Tourism Management 21(1):53-64

21. Prideaux B,Kinin Mount L.(1999).Tourism and Heritage are not strangers:A case study of Opportunities for rural Heritage Museums to maximixe tourism visitation.Journal of Tourism Recreation 37(1):299-303..

22. Richards. G ., (2000) Tourism and the World of Culture and Heritage. Tourism Recreation Research 25(1) 9-17.

23. Rojek .C.,(1993).Ways of Escape :Modern Transformations in Leisure and Travel. The MacMillan Press Ltd; London.

24. Seaton A.V. (1996). Guided by the dark: from thanatopsis to Thana tourism. International Journal of Heritage Studies 2(4):234-244.

25. Sirgy, J. M. (1984). A Social Cognition Model of Consumer Satisfaction/ Dissatisfaction. Psychology and Marketing, 1 (summer), 27-44. 62

26. Timonthy & Wall G.1995. Tourist Accomodation in a Asian historic City.Journal of Tourism Studies. 6(2):63-73.

27. Waitt .G. (2000).Consuming heritage: perceived historical authenticity. Annals of Tourism Research 27(4):835-862

4

EVALUATING STUDY FOR PERFORMANCE OF ROYAL JORDANIAN AND KING HUSSEIN AIRPORT IN LIGHT OF IATA NEW TRENDS

Dr. Farouk Abdelnaby Hassanein Attaallah

Faculty of Tourism and Hotels, Fayoum University, Fayoum, Egypt.

ABSTRACT

The aims of study are to shed light on the most recent trends and issues addressed by the versions of IATA organization for air transport during the period October 2007 - October 2012; and to assess of the current situation in the Royal Jordanian and King Hussein Airport in the light of recent trends of the IATA during the period October 2007 - October 2012. The study tried to examine the Hypotheses of study saying that: the staff of Royal Jordanian Airline and King Hussein Airport is aware of issues in IATA published materials through the period October 2007 – October 2012; the Royal Jordanian Airline and King Hussein Airport apply IATA international standards in the services provided; and there are no significant differences among Sub-Means of the company variable at level 5 %. The Methodology of study consists of Qualitative Approach, Case Study Techniques, Purposive Sample, and SPSS – V – 17. There are limitations of study concluded in scientific limit: IATA versions from October 2007 to October 2012; location limit: Royal Jordanian and King Hussein Airport in Aqaba; and achievement limit: from August 2012 to December 2012. The findings of study reveal that the majority of staff in the Royal Jordanian Airlines and Aqaba Airports Company in Aqaba are unaware of recent issues in IATA published materials. As a result, the hypotheses No. 1 and No. 2 are refused. While, the statistical findings point out that there are no significant differences among sum-means of company variables and the modest value of correlation between the company variable and respondent variable. Consequently and as a whole, the null hypothesis No. 3 is accepted. The study recommends raising the awareness of IATA issues and standards in different areas of King Hussein Airport in Aqaba and the Royal Jordanian Airlines in general and particularly in Aqaba. Also, the importance of conducting more studies to evaluate the range of adopting and application of IATA standards and criteria in airports and airlines performance.

KEYWORDS

Royal Jordanian, King Hussein Airport, Simplifying the Business, Safety, Solar Airplane

1. INTRODUCTION

1.1. Background

The head of the Board of Commissioners of Aqaba Special Economic Zone Authority (ASEZA) Kamel Mohaidein outlined importance King Hussein International Airport in strengthening the investment climate in in Aqaba to become a competing services area in the Middle East region. He noted that it includes a range of investments in the field of air transport more than 100 million dinars (Al Kalaa, 2012, p: 5). ASEZA is keen to raise the efficiency of operation of the airport in accordance with the best international standards to keep pace with global progress in the air transport industry. It will open a group of five-star hotels at the airport, so the airport will be ready to deal with flights and travelers in unprecedented numbers (Al Kalaa, 2012, p: 5).

The general manager of Aqaba Development Company (ADC), Engineer Munir Assad, touches on the strategic plan being implemented to develop airport facilities through 2030 and that the total cost up to nearly 60 million financed by ADC owner of the assets of the airport (Al Kalaa, 2012, p: 5).

The importance of this research is revealed from dealing with a crucial theoretical issue " new IATA trends" in air transport industry being one of the most dynamic economic sector in the world. Moreover, the current study focused on the Royal Jordanian Airline and King Hussein Airport since they have practical value in Jordan.

1.2. Research Questions

The study tries to answer the following two questions:

- Do staff of Royal Jordanian Airline and King Hussein Airport is aware of issues in IATA published materials through period October 2007 – October 2012?.
- Do the Royal Jordanian Airline and King Hussein Airport apply IATA international standards in the services provided?

1.3. Objectives of Study

- Highlighting the most recent trends and issues addressed by the versions of IATA organization for air transport during the period October 2007
- October 2012. - Assessment of the current situation in the Royal Jordanian and King Hussein Airport in the light of recent trends of the IATA during the period October 2007 - October 2012.

2. REVIEW OF LITERATURE

The data have been gathered over time and be organized chronologically from October 2007 to October 2012.

2.1. IATA's STB programme

IATA's Simplifying the Business (STB) programme consists of five projects: electronic ticketing, common use self-service check-in, bar coded boarding passes, radio frequency identification (RFID) for aviation and IATA e-freight (IATA, 2007-13, p: 2).

The programme exceeded all of its board targets for 2008 thanks to its global network of airline and airport managers, solution providers, industry experts and IATA staff as being clear in table no.1. This fact that IATA have exceeded all our board targets shows the importance with which the industry views this programme.

Table 1. Target and actual of STB programme

Project	2008 Target	Actual – Dec 31
ET	100% ET 1 June	Complete
BCBP	200 airlines BCBP capable	204 airlines BCBP capable
CUSS	130 airports offering CUSS	135 airports offering CUSS
IATA e-freight	14 live locations	18 live locations
BIP	Launch at 6 airlines, 6 airports	Launch at 9 airlines, 9 airports
Fast Travel	12 pilots, 2 in each of 6 areas	18 pilots

2.1.1. Useful postings on the StB Support Portal: Interactive Maps

IATA presented a series of Interactive Maps using Google Maps offering useful information by airport to track ET, BCBP and CUSS implementation around the world. The three maps cover more than 1000 airports and will be updated on a monthly basis to provide the industry with the latest view of airport progress (IATA, 2007-19, pp: 2-3).

2.1.2. CUSS implementations around the world

Specific information about their CUSS implementation in live airports include (IATA, 2007-19, p: 3):

- Date CUSS was first installed
- Kiosk vendor
- Number of airlines currently using the kiosks
- Charges
- Bag drop availability, and more.

Air experts expected that travel experience with expanded self-service options give passengers more control over their journeys. The new strategy is built around the success of the Common-Use Self-Service Kiosk, already operating at 83 airports around the world. Moreover, baggage management will help mitigate the US$3 billion in annual costs from the 1.8% of bags that are mishandled (IATA, 2007-20, p: 2). More than half of all passengers surveyed in the latest IATA Corporate Air Travel Survey in 2010 want airlines to provide them with more self-service options. The Fast Travel Program focuses on five areas in a passenger's journey where self-service can be introduced: bags ready-to-go (bag registration), document check, Flight re-booking, self-boarding and bag recovery 'lost bag registration' (IATA, 2010- 50, p: 3).

2.1.3. Electronic ticketing – ET

On 27 August 2007, The International Air Transport Association (IATA) announced it has placed its final order for paper tickets. And it was estimated in just 278 more days the paper ticket will become a collector's item (IATA, 2007-5, p: 1). Airlines should be able to manage the transition to full electronic ticketing on 1 June 2008 (IATA, 2007-10, p: 2).

In order to ensure a smooth transition to a 100% ET world on 1 June 2008, IATA has forwarded to airlines and travel agents using IATA BSPs documentation covering key issues such as (IATA, 2007-25, p: 3):

- Use of the Virtual Miscellaneous Document (V-MPD) and Electronic Miscellaneous Document (EMD).

- .Consequences of a neutral paper ticket being issued on/after 1 June 2008

- Impact of the elimination of neutral paper tickets on 3rd parties in the BSP

There are some primary challenges concerning ET: ground handling, many airlines have not yet implemented a solution for their ground handler to check in passengers with etickets and as a result airlines need to find immediate alternatives rather than waiting for their ground handler to develop the ideal solution; Airlines also need to think about alternative ticketing methods for those tickets that agents will not be able to issue as ETs from 1st of June 2008. Alternatives include placing their own airline tickets in travel agencies or, more likely, having agents issue Multi Purpose Documents (MPDs) or virtual MPDs (IATA, 2008- 8, p: 2).

2.1.4. Bar coded boarding passes - BCBP

Mobile phone check-in enables airlines to send 2D bar codes directly to a passenger's mobile phone, personal digital assistant or smart phone. Passengers simply register their mobile number with their airline at the time of booking to receive a text message with a 2D bar code, or instructions to download it. The bar code becomes the passenger's boarding pass and it is read directly from the screen of the mobile device, eliminating paper completely from the check-in process (IATA, 2007-13, p: 1).

The industry has set a deadline of the end of 2010 to implement 100% bar coded boarding passes (BCBP). Upon full implementation, BCBP will save the industry over US$500 million annually. The global introduction of BCBP to replace magnetic stripe technology is one of five simplifying the Business (StB) projects launched by IATA in 2004 (IATA, 2007-13, p: 2).

According to the Joint Passenger Services Conference (JPSC), the year 2010 is sunset deadline for mag stripe encoded boarding passes into the ticketing resolutions, thereby formally signaling the end to mag stripe encoded ATB2 boarding passes. Carriers will now be able to include a

2D bar code in the ET itinerary receipt enabling passengers to get a quick and direct access to the check-in transaction at the kiosk. By scanning the bar code on the itinerary receipt the kiosk application will identify the passenger and its flight and immediately show the corresponding flight details and options. The passenger is then only one click away from his/her (bar coded) boarding pass. This was possible through changes/additions to Resolutions 722f (IATA, 2007-15, p: 2).

To help airlines and airports match their BCBP implementation plans, IATA has pleased to announce the launch of the BCBP Matchmaker. The BCBP Matchmaker is a free and automated service which allows airlines and airports to send each other request for implementation dates (IATA, 2008-8, p: 4)

At the end of 2010, airlines worldwide completed the implementation of the 2D bar coded boarding pass (IATA, 2010-2, p: 2).

2.1.5. IATA e-freight

Paper-free cargo was estimated to become a reality on key trade routes connecting Canada, Hong Kong, the Netherlands, Singapore, Sweden and the U.K. At each location cargo experts from participating airlines, freight forwarders, ground handling agents, local customs administrations and airport authorities worked together closely over 10 months to prepare the pilots (IATA, 2007-16, p: 1).

There are 4 project streams that support the delivery of e-freight (IATA, 2007-16, p: 2):

- Planning and Development
- Implementation
- Data Quality and Technology Management
- Processes and Standards

The electronic airway bill (E-AWB) is one of the 20 documents planned for conversion with e-freight. An industry standard e-AWB that will eliminate the need to print, handle and archive paper airway bills is being trialed by three airlines and 11 freight forwarders. In addition to greater efficiency, the trials are demonstrating that e-AWB can deliver faster air freight shipments with reduced delays because the electronic documents cannot be lost or misplaced (IATA, 2010-10, p: 2).

E-freight penetration stood at 11% at the end of 2011, ahead of the 10% target set by the IATA Board of Governors. IATA sees three components to achieving 100% by 2015. First, we need to understand e-freight is a supply chain initiative driven forward by the Global Air Cargo Advisory Group (GACAG). Second airlines must drive forward the implementation of the e-air waybill (E-AWB). Cathay Pacific and Emirates have led the way with mandating 100% E-AWB in their home markets. And finally IATA must ensure that the rapidly developing BRICS countries are on board. Current E-AWB penetration is 4.6%. IATA is targeting 15% E-AWB penetration by the end of 2012 and 100% by 2014 (IATA, 2012-17, p: 2).

- **The Baggage Management Improvement Programme**

This aims to ensure the passenger is reunited with their baggage at final destination by improving baggage processes to reduce baggage mishandling. IATA will provide an industry methodology to reduce baggage mishandling through a solutions toolkit, workshops and an internet tool to facilitate cooperation among industry stakeholders. In 2008, IATA secured the participation of 6 airlines and airports in the programme (IATA, 2007- 22, p: 2).

The year 2010 saw IATA bring the total number of airports visited as part of the Baggage Improvement Program (BIP) up to 46. The total number of airports participating in the selfhelp program went up to 27. The most valuable solution in the IATA BIP toolkit is the introduction of baggage tracking (IATA, 2011- 7, p: 3).

IATA now counting down to the end of the Baggage Improvement Program (BIP) which has to be concluded in December 2012 (IATA, 2012- 6, p: 3).

2.2. Environmental Performance

Air Transport industry efficiency translates into improved environmental performance. Airlines contribute 2% of manmade C02. IATA's vision is to do even better. It was targeted carbon neutral growth in the near term. And in the longer-term the goal is nothing less than to become carbon-free. The challenge for the 190 contracting States of

ICAO is to deliver the global political leadership needed to bring this vision to reality (IATA, 2007-7, p: 2).

2.2.1. IATA's Environment Programme

According to Paul Steele, the organisation's Chief Operating Officer for IATA, air transport must be actively engaged in building a sustainable future through many environmental opportunities. IATA's vision is based on a four-pillar strategy (IATA, 2007-17, p: 1):

- Investment in new technology.
- Environmentally efficient infrastructure.
- Improved operations based on industry best practice.
- Economic measures that provide effective incentives to improve environmental performance.

2.2.2 Aviation and Climate Change

All 190 ICAO contracting states have re-affirmed ICAO's leadership on aviation and environment by endorsing a comprehensive strategy on climate change, and embraced IATA's 25% goal to improve fuel efficiency by 2020 (IATA, 2007-11, p: 1). While the political focus has been on economic measures, It is clear that emissions trading alone is not a solution for tackling climate change. CO2 emissions are directly related to fuel consumption. With the fuel bill at 28 % of airline operating costs, this industry is already the most highly incentivised to improve its environmental performance within the limits of technology (IATA, 2007-11, p: 2).

IATA has aligned the industry with a four pillar strategy to address climate change: (1) invest in new technology, (2) fly planes effectively, (3) build and operate efficient infrastructure and (4) call for positive economic incentives to encourage improved fuel efficiency and a reduction in CO2 emissions (IATA, 2008-2, p: 2).

Reducing fuel burn is at the heart of the issue. Every litre of fuel we can save reduces CO2 emissions by over 3 kilograms. Shortening routes, improving air traffic management and spreading best practice in fuel conservation are all approaches to this achievement (IATA, 2008-5, p: 1).

IATA has taken the initiative to supply high-resolution imagery to its member airlines. These initiatives will help airlines to develop fuel-efficient procedures that will reduce their C02 footprint and will enhance safety margins, improve take-off and landing frequency and thus shorten gate-to-gate times (IATA, 2008-27, p: 5).

Our vision for carbon neutral growth leading to a carbon-free future sets the benchmark (IATA, 2008- 33, p: 2). According to Samer Majali, Royal Jordanian Airlines CEO, the incentive for airlines to improve performance has never been bigger, but governments stand in our way. They must overcome their obsession with punitive and unilateral emissions trading schemes and start working on real solutions to reduce CO2 (IATA, 2008-35, p: 5).

Giovanni Bisignani, IATA's Director General and CEO, identified three critical areas that can help the industry deliver even better results (IATA, 2009-5, p: 2):

- Alternative fuels: Bio-fuels show the most promise for reducing aviation's carbon emissions. IATA is committed to using 10% alternative fuels by 2017. But we need the right bio-fuels, those that don't compete with food for land-use or harm bio-diversity, and that meet the current exacting technical specifications of jet kerosene. The recent tests by Air New Zealand and Continental Airlines proved that bio-fuels are viable. Now we need to speed-up the certification process. The current timeline sees certification by 2013.

- Better air navigation: "We need to fly more effectively. Every Continuous Descent Approach (CDA) saves between 150 to 600kg of CO2. Each Clean Airspeed Departure (CAD) saves between 600 to 5,000 kg of CO2.

- A global solution on positive economic measures: Governments around the world must agree a global solution to reduce aviation emissions and refuses unilateral approach of Emissions Trading Scheme (ETS). The aviation industry is united behind three targets: (1) a 1.5% average annual improvement in fuel efficiency to 2020, (2) capping net emissions from 2020 with carbonneutral growth and (3) cutting emission in half by 2050 compared to 2005 (IATA, 2010-71, p: 1):

The aviation industry is united behind three targets: (1) a 1.5% average annual improvement in fuel efficiency to 2020, (2) capping net emissions from 2020 with carbonneutral growth and (3) cutting emission in half by 2050 compared to 2005 (IATA, 2010-71, p: 1)

2.2.3. Solar airplane

IATA has become an Institutional Partner of Solar Impulse– the solar airplane that will fly around the world with no fuel and zero emissions. The agreement, signed by Bertrand Piccard, President of Solar Impulse and Giovanni Bisignani, Director General and CEO of IATA, establishes a framework for cooperation between the two organizations. IATA will provide assistance to ensure the smooth passage of the solar plane around the world (IATA, 2008-4, p: 1).

2.3. Safety

IOSA plays an important role. 136 airlines are on the IATA Operational Safety Audit (IOSA) registry but cargo is behind with only one airline on the registry. The goal is to bring all carriers on board by the end of the year 2008 (IATA, 2008-9, p: 1). Cooperation and a broader view based on global standards are necessary to address unique infrastructure challenges and upgrade the ageing and insufficient air traffic control technology (IATA, 2008-15, p: 2).

Flight Data Analysis (FDA) is a tool of IOSA. FDA offers airlines the capability to statistically analyse data from actual flights to improve procedures and monitor compliance. FDA offers critical insight for aircraft maintenance by using data to monitor engine condition trends and aircraft fuel usage (IATA, 2008-16, p: 2).

IATA's three other key areas to improve safety are (IATA, 2008-16, p: 2):

- Auditing: e.g. Egypt is mandating IOSA as part of their safety oversight programmes.

- Infrastructure: IATA highlighted two areas of concern. Despite high user charges, in many airports, infrastructure is poorly funded and not up to international standards. Lack of transparency is a critical issue that is costing lives. IATA supports the creation of special infrastructure fund mechanisms to ensure that the money that airlines pay in charges stays in the industry.

- Skills Shortage: IATA expected the shortfall of skilful staff in air industry by 2026 and it broadens the pool of qualified candidates without compromising on safety.

Almost half (48%) of the year's accidents took place during landing. The majority of these accidents involved a runway excursion. Many of these accidents could have been prevented by the initiation of a timely go-around. IATA, in cooperation with the Flight Safety Foundation, is developing a toolkit that will address the issues linked to runway safety enhancement, including the prevention of runway excursions. Nearly 20% of accidents 2007 related to ground damage. Lack of standardisation can contribute to ground handling activities that result in damage to aircraft. IATA developed the IATA Safety Audit for Ground Operations (ISAGO) programme to drastically reduce aircraft damage and personal injuries in the ground environment. About half of the accidents in 2007 were linked to a technical issue; maintenance events contributed to almost 20% of all occurrences last year. IATA is revising its safety strategy to encompass maintenance activities and Safety Management System implementation for maintenance organisations (IATA, 2008-24, p: 2):

As of 1 January 2009, IOSA has been a condition of IATA membership. Airlines that have not passed the quality control process by 31 March 2009 will have their memberships terminated (IATA, 2009-12, p: 1):

Safety is a constant challenge. Having made aviation the safest way to travel, further improvements will come only with careful data analysis. We must understand the underlying safety risk trends, not just from the handful of accidents each year, but by bringing together and analyzing data from millions of safe flights. The IATA Global Safety Information Center was launched in December 2009 for just that purpose (IATA, 2010-6, p: 2):

IATA is eager to work with governments to make IOSA a part of their safety oversight programs in air transport. IATA conducted accident analysis in 2011 and results were ((IATA, 2012-15, p: 2) :

- Runway excursions, in which an aircraft departs a runway during a landing or takeoff, were the most common type of accident in 2011 (18% of total accidents). Unstable approaches-- situations where the aircraft is too fast, above the glide slope, or touches down beyond the desired touchdown point--and contaminated runways are among the most common contributing factors to runway excursions on landing.

- IATA's Global Safety Information Center (GSIC) provides trend analysis that is helping the industry improve performance. For example, a new Flight Data exchange (FDX) system within the GSIC tracks unstable approach performance for the more than 700 airports in the database. Sharing such safety data complements the work of the Runway Excursion Risk Reduction (RERR) Toolkit, the second edition of which was launched in May 2011, and fuels global efforts to find solutions.

- Ground damage, accounting for 16% of accidents in 2011. This was up from 11% in 2010. These accidents include events such as damage resulting from ground handling operations and collisions during taxi. IATA has launched a number of initiatives to address ground accidents. In 2008, IATA launched the IATA Safety Audit for Ground Operations (ISAGO). This is the first global standard for the oversight and auditing of ground handling companies. Subsequently, IATA has launched the IATA Ground Operations Manual (IGOM) and the Aircraft Ground Damage Database (GDDB) to collect and analyze reports of ground damage from participating operators and ground service providers.

- Sharing information, data sharing is helping to identify and reduce risks. In 2010 the International Civil Aviation Organization, the US Department of Transportation and the European Commission agreed with IATA to create the Global Safety Information Exchange (GSIE). This was enriched in 2011 with the addition of databases covering ground damage and flight data.

➢ **IATA DGR e-newsletter**

The objective of IATA DGR e-newsletter is to provide users of the Dangerous Goods Regulations (DGR) with a better understanding of how to interpret and apply the Dangerous Goods Regulations (IATA, 2007-9, p: 1).

2.4. Security

IATA has worked on internationally recognized security accreditation standards and a security audit. This includes a global registry of secure supply chain operators, including known shipper details to be used by regulators and the supply chain (IATA, 2008-9, p: 2).

The IATA Training and Development Institute ITDI's Diploma in Aviation Security is designed to ensure participants have an opportunity to learn about every aspect of security and facilitation (IATA, 2008-27, p: 3).

Security is an uncoordinated mess. Since 2001 till 2008 airlines and their customers nearly have paid over US$30 billion for security measures. Governments must do their part and focus on risk management, harmonise global standards, make better use of technology, and take responsibility for the bill (IATA, 2008-33, p: 2).

Governments and industry are aligned in the priority that we place on security. But the cost of security is also an issue. Globally, airlines spend US$5.9 billion a year on what are essentially measures concerned with national security ((IATA, 2010-2, p: 2).

2.5. Efficiency and quality

Cargo 2000 was established over a decade ago to simplify processes and implement effective quality standards from order to delivery. IATA simplified the supply chain process from 40 to 19 steps and developed common parameters for quality measurement. Moreover, IATA quality audit based on Cargo 2000 standards is another support in this area (IATA, 2008-9, p: 2).

➢ **Airport Infrastructure**

The airport structure and regulatory system play a critical role to deliver world-class airport facilities. Lack of investment in an airport infrastructure delivers embarrassingly-low service levels on everything from security wait-times to baggage delivery and almost everything in-between (IATA, 2007-4, p: 1).

IATA, along with the majority of other industry groups, support the use of IATA's Worldwide Scheduling Guidelines (WSG) to manage congestion at international airports (IATA, 2007-24, p: 1).

3. METHODOLOGY

In this section the researcher gives details about the method of research used and analysis of the information gathered for the current study.

3.1. The Pilot Study

The researcher conducted a pilot study adopting brainstorm techniques of his colleagues about: the title of study, the questions of research, methodology and questionnaire design.

3.2. Descriptive Analytical Method and Case Study Technique

Descriptive Analytical Approach handles the current phenomenon as it is in the real field, whether by a qualitative technique; case study technique. The principal purpose of the descriptive method is to evaluate things and conditions in their natural cases. Regarding the case study method, it has the crux importance in data collection and analysis in Descriptive Approach.

3.3. SPSS for Data Analysis

For techniques of analysis, statistics models that are suitable for measuring and analyzing attitudes and opinions towards a phenomenon are the Frequency, the Mean, the Standard Deviation, the Standard Error of Mean, Test of Variance, ETA Value of Correlation, and ETA Square Value of Regression.

3.4. Sample and Questionnaire

The current research depends on the purposive sample for data collection. The purposive sample includes subjects selected on the basis of specific characteristics or qualities. In a similar method, the quota sample, subjects are selected to meet a predetermined or known percentage.

The purposive sample of the current study includes staff of Royal Jordanian Airlines (3 forms), and staff of Aqaba Airports Company working at King Hussein Airport (7 forms). The current study depends on intensive interviews, or in-depth interviews, which are essentially a hybrid of the one-on-one personal interview approach. All distributed forms of questionnaire are answered and returned back.

Furthermore, this study adopted the Likert Scale, with choices strongly disagree = 1, disagree = 2, neutral = 3, agree = 4 and strongly agree = 5; and the Dichotomous Choice, with choices Yes = 1 and No = 2; since they are preferable scales to measure attitudes and points of view towards a specific issue or phenomenon (appendix no. 2).

Table 2. Results of respondents according to background of IATA site

Item	Frequencies		Mean	Std.Error of Mean	Std. Deviation	F Sig.	ETA	ETA Square
	Yes	No						
Background of IATA site	54.5%	45.5%	1.45	0.157	0.522	0.662	0.149	0.022

4. RESULTS AND DISCUSSION

Regarding the table no. 2 mentioned above, we can say that most of respondents have no background about information of IATA's site; since the Mean degree is 1.45 with no dispersion among answers of respondents (Std. Deviation = 0.522) and no dispersion from the Mean of population (Std. Error of Mean = 0.157). Moreover, there are no significant differences among Sub-Means of company variable (F. Significance = 0.662 < 0.05). Consequently, we accept the null hypothesis. In additions, there is weak correlation between the company variable and respondent variable (ETA = 0.149), and the area variable has less impact on the respondent variable (ETA Square = 0.022). One respondent mentions that his background on IATA site is determined in three areas: areas of activities, events, and training. Another one has background just on training programs, since he had training on terminal buildings planning standards (LOS).

Table 3. Results of respondents concerning with Background of STB.

Item	Frequencies		Mean	Std Error of Mean	Std. Deviation	F Sig.	ETA	ETA Square
	Yes	No						
Background of STB	63.6%	36.4%	1.36	0.152	0.505	0.910	0.039	0.001

Regarding the table no. 3, we can say that plurality of respondents have no background about information of IATA's Simplifying Business Program; the Mean degree is 1.36 with no dispersion among answers of respondents (Std. Deviation = 0.505) and no dispersion from the Mean of population (Std. Error of Mean = 0.152). Moreover, there are no significant differences among Sub-Means of company variable (F. Significance = 0.910 < 0.05). Accordingly, we accept the null hypothesis. Moreover, there is little correlation between the company variable and respondent variable (ETA = 0.039), and the company variable has feeble impact on the respondent variable (ETA Square = 0.001). Some of respondents comments that his background about the STB program is concluded in electronic ticketing and common use self-service check-in. Another respondent points out that the STB program has e-freight, baggage improvement programs, fast travel IATA e-services, passenger facilities, and new distribution capabilities

Table 4. Results of respondents pertaining to application of STB

Item	Frequencies						Mean	Std.Error of Mean	Std. Deviation	F Sig	ETA	ETA Square
	S Agree	Agree	Neutral	Disagree	S. Disagree	Total						
Application of STB	—	27.3%	45.5%	27.2%	—	100%	3.00	0.234	0.775	0.078	0553	0.306

Given the table no. 4 , arguably, preponderance of respondents are not able to decide about the application of IATA's Simplifying Business Program; the Mean degree is 3.00 with no dispersion among answers of respondents (Std. Deviation = 0.775) and no dispersion from the Mean of population (Std. Error of Mean = 0.234). Moreover, there are no significant differences among Sub-Means of company variable (F.

Significance = 0.078 < 0.05). As a result, we accept the null hypothesis. Furthermore, there is moderate correlation between the company variable and respondent variable (ETA = 0.553), and the company variable explains in an assuming manner the respondent variable (ETA Square = 0.306). One respondent states that STB program is applied only in sales and reservation and check in service.

Table 5. Results of respondents regarding to use of IATA interactive maps

Item	Frequencies						Mean	Std.Error of Mean	Std. Deviation	F Sig.	ETA	ETA Square
	S. Agree	Agree	Neutral	Disagree	S. Disagree	Total						
Use IATA interactive maps	—	27.3%	54.5%	18.2%	—	100%	2.91	0.211	0.701	0.237	0.389	0.151

Seeing that the table no. 5, it may be said that majority of respondents also can't give a specific decision about the using the IATA interactive maps in real practices or not; the Mean degree is 2.91 with no dispersion among answers of respondents (Std. Deviation = 0.701) and no dispersion from the Mean of population (Std. Error of Mean = 0.211). Moreover, there are no significant differences among Sub-Means of company variable (F. Significance = 0.237 < 0.05). As a result, we accept the null hypothesis. Furthermore, there is modest correlation between the company variable and respondent variable (ETA = 0.389), and the company variable explores in a little manner the respondent variable (ETA Square = 0.151).

Table 6. Results of respondents regarding to Background on IATA environmental program

Item	Frequencies		Mean	Std.Error of Mean	Std. Deviation	F Sig.	ETA	ETA Square
	Yes	No						
Background on IATA environmental program	45.5%	54.5%	1.55	0.157	0.522	0.662	0.149	0.022

Seeing that the table no. 6, we can say that plurality of respondents have no idea on the IATA's environmental program; the Mean degree is 1.55 with no dispersion among answers of respondents (Std. Deviation = 0.522) and no dispersion from the Mean of population (Std. Error of

Mean = 0.157). In additions, there are no significant differences among Sub-Means of company variable (F. Significance = 0.662 < 0.05). Consequently, we accept the null hypothesis. Furthermore, there is humble correlation between the company variable and respondent variable (ETA = 0.149), and the company variable clears out in a slight manner the respondent variable (ETA Square = 0.022).

Table 7. Results of respondents pertaining to application of IATA environmental program

Item	Frequencies						Mean	Std Error of Mean	Std. Deviation	F Sig.	ETA	ETA Square
	S. Agree	Agree	Neutral	Disagree	S. Disagree	Total						
Application of IATA environmental program	—	27.3%	54.5%	18.2%	—	100%	4.00	0.234	0.775	0.411	0.276	0.076

Regarding the table no. 7, arguably that preponderance of respondents also agree that there is application of IATA environmental programme in real field to some range; the Mean degree is 4.00 with no dispersion among answers of respondents (Std. Deviation = 0.775) and no dispersion from the Mean of population (Std. Error of Mean = 0.234). This is contrast to the results in the previous point concerning the background on IATA's environmental programme. Moreover, there are no significant differences among Sub-Means of company variable (F. Significance = 0.411 < 0.05). As a result, we accept the null hypothesis. Furthermore, there is modest correlation between the company variable and respondent variable (ETA = 0.276), and the company variable explores in a negligible manner the respondent variable (ETA Square = 0.076). Some of respondents say that IATA's environmental programme is applied in solid waste management.

Table 8. Results of respondents regarding to Background on solar airplane

Item	Frequencies		Mean	Std.Error of Mean	Std. Deviation	F Sig.	ETA	ETA Square
	Yes	No						
Background on solar airplane	36.4%	63.6%	1.64	0.152	0.505	0.241	0.386	0.149

Seeing that the table no. 8, we can say that plurality of respondents has no background on the solar airplane; the Mean degree is 1.64 with no dispersion among answers of respondents (Std. Deviation = 0.505) and no dispersion from the Mean of population (Std. Error of Mean = 0.152). Moreover, there are no significant differences among Sub-Means of company variable (F. Significance = 0.241 < 0.05). Accordingly, we accept the null hypothesis. Furthermore, there is slight correlation between the company variable and respondent variable (ETA = 0.386), and the company variable reveals in a little manner the respondent variable (ETA Square = 0.149). One respondent comments that it is too early that the solar airplane flies around the world.

Table 9. Results of respondents regarding to adoptation of solar airplane

Item	Frequencies						Mean	Std.Error of Mean	Std. Deviation	F Sig.	ETA	ETA Square
	S. Agree	Agree	Neutral	Disagree	S. Disagree	Total						
Adoptation of solar airplane	9.1%	18.2%	72.7%	—	—	100%	2.64	0.203	0.674	0.389	0.289	0.083

Giving in the table no. 9, we can say that preponderance of respondents is neutral to decide about adoptation of solar airplane in air transport business herein; the Mean degree is 2.64 with no dispersion among answers of respondents (Std. Deviation = 0.674) and no dispersion from the Mean of population (Std. Error of Mean = 0.203). Moreover, there are no significant differences among Sub-Means of company variable (F. Significance = 0.389 < 0.05). Accordingly, we accept the null hypothesis. Furthermore, there is slight correlation between the company variable and respondent variable (ETA = 0.289), and the company variable reveals in a trifle manner the respondent variable (ETA Square = 0.083).

Table 10. Results of respondents pertaining to Background on areas of safety

Item	Frequencies		Mean	Std.Error of Mean	Std. Deviation	F Sig.	ETA	ETA Square
	Yes	No						
Background on areas of safety	54.5%	45.5%	1.45	0.157	0.522	0.662	0.149	0.022

Seeing that the table no. 10, we can say that majority of respondents has no background on areas of safety; the Mean degree is 2.64 with no dispersion among answers of respondents (Std. Deviation = 0.522) and no dispersion from the Mean of population (Std. Error of Mean = 0.157). Moreover, there are no significant differences among Sub-Means of company variable (F. Significance = 0.662 < 0.05). As a result, we accept the null hypothesis. Furthermore, there is slight correlation between the company variable and respondent variable (ETA = 0.149), and the company variable reveals in a trifle manner the respondent variable (ETA Square = 0.022). One respondent clears out that his background about safety in air transport is gained from ICAO.

Table 11. Results of respondents concerning to application of keys of safety

Item	Frequencies						Mean	Std Error of Mean	Std. Deviation	F Sig	ETA	ETA Square
	S. Agree	Agree	Neutral	Disagree	S. Disagree	Total						
Application of keys of safety	18.2%	18.2%	54.5%	—	9.1%	100%	3.36	0.338	1.120	0.538	0.209	0.043

Giving in the table no. 11, we can say that preponderance of respondents is agree that there is application of keys of safety in real performance; the Mean degree is 3.36 with dispersion among answers of respondents (Std. Deviation = 1.120) and no dispersion from the Mean of population (Std. Error of Mean = 0.338). This is contrast to the results in the previous point concerning the background on IATA's areas of safety. Moreover, there are no significant differences among Sub-Means of company variable (F. Significance = 0.538 < 0.05). Accordingly, we accept the null hypothesis. Furthermore, there is slight correlation between the company variable and respondent variable (ETA = 0.209), and the company variable clears out in a modest manner the respondent variable (ETA Square = 0.043). Respondents of AAC say that they always attend airport safety and security inside and outside the KHA , which are conducted by ICAO certified trainers.

Table 12. Results of respondents regarding to background on FDA

Item	Frequencies		Mean	Std.Error of Mean	Std. Deviation	F Sig	ETA	ETA Square
	Yes	No						
Background on FDA	27.3%	72.7%	1.73	0.141	0.467	0.256	0.375	0.141

Given in the table no. 12, we can say that majority of respondents has no background on FDA; the Mean degree is 1.73 with no dispersion among answers of respondents (Std. Deviation = 0.467) and no dispersion from the Mean of population (Std. Error of Mean = 0.141). In additions, there are no significant differences among Sub-Means of company variable (F. Significance = 0.256 < 0.05). As a result, we accept the null hypothesis. Furthermore, there is weakly correlation between the company variable and respondent variable (ETA = 0.375), and the company variable reveals in a slight manner the respondent variable (ETA Square = 0.141).

Table 13. Results of respondents concerning to application of FDA

Item	Frequencies						Mean	Std.Error of Mean	Std. Deviation	F Sig.	ETA	ETA Square
	S. Agree	Agree	Neutral	Disagree	S. Disagree	Total						
Application of FDA	—	27.3%	45.5%	18.2%	9.1%	100%	2.91	0.285	0.944	0.233	0.392	.0.153

Regarding the table no. 13, we can say that preponderance of respondents is neutral about the application of FDA in real performance; the Mean degree is 2.91 with dispersion among answers of respondents (Std. Deviation = 0.944) and no dispersion from the Mean of population (Std. Error of Mean = 0.285). Moreover, there are no significant differences among Sub-Means of company variable (F. Significance = 0.233 < 0.05). Accordingly, we accept the null hypothesis. Furthermore, there is slight correlation between the company variable and respondent variable (ETA = 0.392), and the company variable clears out in a modest manner the respondent variable (ETA Square = 0.153).

Table 14. Results of respondents regarding to background on diploma of TTDI

Item	Frequencies		Mean	Std.Error of Mean	Std. Deviation	F Sig.	ETA	ETA Square
	Yes	No						
Background on diploma of ITDI	63.6%	36.4%	1.36	0.152	0.505	0.910	0.039	0.001

Given in the table no. 14, we can say that majority of respondents has no background on diploma of ITDI; the Mean degree is 1.36 with no dispersion among answers of respondents (Std. Deviation = 0.505) and no dispersion from the Mean of population (Std. Error of Mean = 0.152). In additions, there are no significant differences among Sub-Means of company variable (F. Significance = 0.910 < 0.05). As a result, we accept the null hypothesis. Furthermore, there is weakly correlation between the company variable and respondent variable (ETA = 0.039), and the company variable reveals in a slight manner the respondent variable (ETA Square = 0.001).

Table 15. Results of respondents regarding to training of staff on diploma of TTDI

Item	Frequencies						Mean	Std.Error of Mean	Std. Deviation	F Sig.	ETA	ETA Square
	S. Agree	Agree	Neutral	Disagree	S. Disagree	Total						
Training of staff on diploma of ITDI	18.2%	18.2%	45.5%	9.1%	9.1%	100%	3.27	0.359	1.191	0.020	0.686	0.471

Seeing that the table no. 15, we can say that preponderance of respondents has the diploma of ITDI; the Mean degree is 3.27 however dispersion among answers of respondents (Std. Deviation = 1.191) and no dispersion from the Mean of population (Std. Error of Mean = 0.359). Moreover, there are significant differences among Sub-Means of company variable (F. Significance = 0.020 > 0.05). Accordingly, we accept the alternative hypothesis. Furthermore, there is moderate correlation between the company variable and respondent variable (ETA = 0.686), and the company variable clears out in a moderate manner the respondent variable (ETA Square = 0.471). Some of respondents explain that the ITDI's Diploma in aviation security is concised for the airport staff only

4. CONCLUSION AND RECOMMENDATIONS

The study achieved its objectives. It demonstrated the mainstay issues of IATA since 2007 to 2012; IATA's STB programme, environmental performance in aviation, Safety, and security. Also, there is an estimation of the current status quo regarding the Royal Jordanian Airlines and King Hussein Airport in Aqaba.

The results of discussion in previous section show that the majority of staff in the Royal Jordanian Airlines and Aqaba Airports Company in Aqaba are unaware of recent issues in IATA published materials through the period from October 2007 to October 2012 as a scientific determinant of this current study. Shortage of staff's knowledge addresses both the background and application of each issue. Accordingly, we refuse hypotheses No. 1 and No. 2.

Moreover, the statistical findings reveal that there are no significant differences among sum-means of company variables and the modest value of correlation between the company variable and respondent variable. Except in one criterion concerning the diploma in the aviation security conducted by the IATA Training and Development Institute. The respondents say that this diploma is pertaining only to the staff of airports and not for those of airlines. Consequently and as a whole, we accept the null hypothesis No. 3.

The study recommends raising the awareness of IATA issues and standards in different areas of King Hussein Airport in Aqaba and the Royal Jordanian Airlines in general and particularly in Aqaba. This research proposes a five-years action plan for adopting and application of IATA standards and criteria in airports and airlines performance including the following steps:

1) Forming steering committee of specialists in the two cases of study.

2) Placing the mission of performance in the two entities.

3) Placing the message of performance in the two entities.

4) Determining of performance's objectives in the two entities.

5) Specifying of performance's strategies to achieve objectives.

6) Choosing techniques to implement strategies.

7) Specifying the areas of performance that need changes and promotion.

8) Adopting standards of best practices in each area.

9) Monitoring. The performance.

10) Continuous Feedback for future improvements.

REFERENCES

1. Al Kalaa Magazine (2012: 131): "The Importance of Aqaba Airport", Halalat Press Company, Aqaba, Jordan.
2. IATA (2007-4): "Fixing the UK's Airports", Montreal, Quebec, Canada.
3. IATA (2007-5): "IATA Issues 'Last Call' for Paper Tickets", Montreal, Quebec, Canada.
4. IATA (2007-7): "Industry Registers Record Traffic Volumes, Load Factors in July ", Montreal, Quebec, Canada.
5. IATA (2007-8): "IATA Partner Brief- Quarter 3, 2007", Montreal, Quebec, Canada.
6. IATA (2007-9): "DGR e-newsletter- September 2007", Montreal, Quebec, Canada.
7. IATA (2007-10): "Simplifying the Business Newsletter- Issue 22", Montreal, Quebec, Canada.
8. IATA (2007-11): "Governments Endorse Global Strategy for Aviation and Climate Change", Montreal, Quebec, Canada.
9. IATA (2007-13): "IATA Standard Paves Way for Global Mobile Phone Check-in ", Montreal, Quebec, Canada.
10. IATA (2007-14): "Passenger numbers to reach 2.75 billion by 2011", Montreal, Quebec, Canada.
11. IATA (2007-15): "Simplifying the Business Newsletter- Issue 23", Montreal, Quebec, Canada.
12. IATA (2007-16): "IATA e-freight pilots go live", Montreal, Quebec, Canada.
13. IATA (2007-17): "New Director Raises Profile of IATA's Environment Programme ", Montreal, Quebec, Canada.

14. IATA (2007-19): "Simplifying the Business Newsletter- Issue 24", Montreal, Quebec, Canada.

15. IATA (2007-20): "New IATA Financial Forecast Predicts 2008 Downturn", Montreal, Quebec, Canada.

16. IATA (2007-20): "StB FLASH newsletter- December 2007", Montreal, Quebec, Canada.

17. IATA (2007-23): "IATA Launches www.iatatravelcentre.com-Information Portal for Travelers", Montreal, Quebec, Canada.

18. IATA (2007-24): "IATA Rejects Slot Auctions for New York", Montreal, Quebec, Canada.

19. IATA (2007-25): "Simplifying the Business Newsletter- Issue 25", Montreal, Quebec, Canada.

20. IATA (2008-2): "Airport Privatization and Environment: Two Leadership Challenges for Japan", Montreal, Quebec, Canada.

21. IATA (2008-4): "IATA joins forces with Solar Impulse", Montreal, Quebec, Canada.

22. IATA (2008-5): "IATA Calls for Environmental Leadership in Asia", Montreal, Quebec, Canada.

23. IATA (2008-8): "Simplifying the Business Newsletter, Issue 27", and Montreal, Quebec, Canada.

24. IATA (2008-9): "Improve Air Cargo Competitiveness", Montreal, Quebec, Canada.

25. IATA (2008-15): "Urgent Work Needed To Improve Latin America Safety", Montreal, Quebec, Canada.

26. IATA (2008-16): "Improving African Safety", Montreal, Quebec, Canada.

27. IATA (2008-17): "IATA Welcomes Shorter Route over China", Montreal, Quebec, Canada.

28. IATA (2008-24): "Press Release: Industry Sees Fewer Fatalities", Montreal, Quebec, Canada.

29. IATA (2008-27): "Safety, Operations and Infrastructure Newsletter-May 2008", Montreal, Quebec, Canada.

30. IATA (2008-33): "Fuel Crisis a Catalyst for Change", Montreal, Quebec, Canada.

31. IATA (2008-35): "Press Release: Royal Jordanian's Majali is New IATA Chairman", Montreal, Quebec, Canada.

32. IATA (2009-5): "IATA reiterates its environmental commitment even in times of crisis", Montreal, Quebec, Canada.

33. IATA (2009-9): "StB January Newsletter: As economy worsens, StB stays on track", Montreal, Quebec, Canada.

34. IATA (2009-12): "More Accidents But Fewer Fatalities in 2008", Montreal, Quebec, Canada.

35. IATA (2009-17): "Opportunities for Malaysian Aviation", Montreal, Quebec, Canada.

36. IATA (2010-2): "2009: Worst Demand Decline in History-Encouraging Year-end Improvements", Montreal, Quebec, Canada.

37. IATA (2010-6): "Aircraft Accident Rate Drops in 2009- Renewed Focus on Training, Data", Montreal, Quebec, Canada.

38. IATA (2010-10): "Aircraft Accident Rate Drops in 2009- Renewed Focus on Training, Data", Montreal, Quebec, Canada.

39. IATA (2010-50): "StB at Check-in 2010 - Las Vegas", Montreal, Quebec, Canada.

40. IATA (2010-71): "Opportunity for a Global Framework on Environment", Montreal, Quebec, Canada.

41. IATA (2011-2): "Aviation Opportunities for Vietnam", Montreal, Quebec, Canada.

42. IATA (2011-3): "E-freight achievements and more", Montreal, Quebec, Canada.

43. IATA (2012-3): "StB Newsletter January 2012", Montreal, Quebec, Canada

44. IATA (2012-15): " Press Release: Global Accident Rate Reaches New Low", Montreal, Quebec, Canada

45. IATA (2012-17): "Strong Partnerships across the Air Cargo Value Chain Safety, Security, Sustainability and Competitiveness Top the Agenda", Montreal, Quebec, Canada.

46. IATA (2012-21): "Rising Oil Prices Reducing Profitability Regional Differences Widen", Montreal, Quebec, Canada.

5

THE IMPACT NUTRITIONAL INFORMATION HAS ON CUSTOMERS' BEHAVIOURS TOWARDS THEIR FOOD CHOICES IN RESTAURANTS.

Jenna Carra Musiker, Dr. Hema Kesa

Faculty of Management, School of Tourism and Hospitality, University of Johannesburg, Johannesburg, Gauteng, South Africa

ABSTRACT

In South Africa weight gain and health problems are on a sharp uphill rise. Very few restaurants offer nutritional information or pictures. This study is of importance as it aims at highlighting the demand for this. It investigated customers' behaviours and attitudes towards the foods they wish to consume. Two Johannesburg restaurants were selected to uncover the reasons as to why people choose specific meals, and how if any the addition of nutritional information, health icons and appealing pictures made a difference to their ordering choices. Furthermore it looked at the connection this had to restaurant sales, and the general population of customers who ate at these restaurants. This determined the needs and demands for nutritional information within restaurants. Additional customer feedback, personal views and opinions from the restaurants were questioned to get an overview of what consumers now days tend to look at when dining out.

1. INTRODUCTION

Over the last decade nutritional trends have altered dramatically and have led to more health concerns and a need to live healthier lifestyles. Nutrition is vital for existence and gives people the nourishment that they need for growth and development. According to (Swart, Sanders & McLachlan, 2008:129-148) the overall nutritional status of South Africa"s population has not improved over the last fourteen years. The past several decades have seen an increasing prevalence of nutrition related conditions such as obesity; consequently, there has been an on growing focus on consumer nutritional education (Nestle & Jacobson, 2000:U.S. Food and Drug Administration, 2004). It is of great importance that restaurateurs begin to include nutritional information as well as more appealing visual food pictures, and health icons, aiding in the health aspects and obesity awareness as well as giving people the nutrition they require. The more people eat out, the lower their quality of diet becomes & the higher the impact on their health (Morrison; 2011:10)

1.1. Background of the Study

Customers eat with their eyes, and that is why food must be desirable, and impress both the eyes and the palate (Magoulas, 2009). The look and presentation of foods affect our tastes, moods and inner comfort levels, this is why restaurant owners need to stimulate taste by creating a meal that tastes great, is healthy and looks amazing. "The menu is the number one tool, and is the purest form of a restaurants strategic marketing plan", (McCall and Lynn, 2008:439-445). Menus drive the process for diners to enter restaurants; it serves as a first impression and can tell a lot about the outlet. Menu design is directly linked to and influences how customers perceive things. The placement of menu items, graphics and descriptions tells and sends messages to the customers about what they should order (Panitz, 2000:81-87). They are intended to promote a perception of value to customers, and helps give more direction to what to choose (McCall and Lynn, 2008:439-445). There is a definite need for restaurants to find better ways of showing information, but at the same time they must avoid overloads and visual clutters. For these reasons they say including nutritional information and pictures on restaurant menus and menu boards or table cards would be more beneficial (Thomas Jr. and Mills, 2006:6-22).

In today"s restaurants, nutritional information is rarely offered, but there is demand as more people are aware of healthier options; people are more conscious about what they choose to eat and this leads to a potential impact of enabling people to be more knowledgeable and health conscious about what they consume

"Recently significant attention has been paid to the epidemic of obesity and other health issues that relate to foods people consume" as stated by (Jones, 2009:344-363). Not enough is known about this information on menus in South Africa, so further insight into how health icons, pictures and nutritional information could influence customers" choices will be investigated in this study by the use of randomized samples. The spotlight focuses on the fact that more people eat out away from home and this impacts their health and weights (Jones, 2009:344-363), and to help determine what could be done to encourage healthier food choices and intakes at restaurants (Jones, 2009:344-363). If the placements of information and icons would make a difference, this study explored peoples" attitudes on these factors to see if they do in fact influence customers" behaviours towards their meal decisions.

1.2. Purpose of the Study

The purpose of this study was to investigate and determine whether placing nutritional information (NI), health icons and pictures within restaurant would influence the customers" behaviours regarding their food choices. The study evaluated the impact of menu labelling in two restaurants. Would this influence customers overall decisions in their meal choices? This was an important question, as individuals must maintain healthier lives, and receive the best nutrients that they require. Today more people are eating away from home, and in order to keep them aware of what they are consuming, restaurant owners should include labelling in their restaurants. By doing so it hoped to assist customers in making better meal choices. The question is would it actually affect their purchasing and consumption patterns?

South Africa has taken a turn in the direction of obesity and other illnesses, and it is largely attached to what individuals are eating. The researchers concern is that if there has been no visible improvement in people"s diets, are our health issues, aimed at restaurants specifically, going to become more of a problem? Another frequent question is; are restaurants serving good enough foods to the public, due to the health issues that are arising? This study investigated these research questions, as it hoped to increase customer satisfaction and improve perception potentially by having information about food items. By bringing menu labels, icons and pictures into restaurants, it is a strategy helped at reducing obesity and diet related chronic diseases, leading to individuals taking more control in their food choices and having healthier lives. It hoped to show that by incorporating these factors into restaurants, the industry can benefit from additional research knowledge on people"s attitudes towards nutritional information and icons displayed, as well as to have created more awareness of nutritional issues in general and sustain healthier lifestyles through the food choices people pick. The purpose also determined whether any changes occurred in sales, as well as if it improved customers" behaviours that could result in repeat patronage. It showed that customers choose foods based on information and new knowledge about health, as well as on how the pictures portray the meals. This paper could help support hospitality service providers/restaurateurs with keys to understanding just how important nutritional information, health icons and visual aids could be.

1.3. Objectives of the Study

The objective of the study was to research the impact of nutritional information, health icons and pictures in restaurants, and see the effects of these on customers" behaviours and choices. In particular the following were assessed:

- Customers" choices and behaviours on their food decisions.
- How both the use of pictures and nutritional information on menus affected the customers" choice in food items?
- Identified whether these changes had an impact on the purchasing patterns to increase or decrease the restaurants sales, as well as the impact on people"s knowledge and future health benefits

The aim of this study was designed to evaluate and report these influences, and if it would assess their effectiveness. To understand nutritional information in restaurants, individuals must be nutritionally aware and educated about healthier options out there.

2. LITERATURE REVIEW

2.1 Obesity and Health Concerns

Dining out used to be reserved for special occasions, whereas recently it is a part of many people"s daily lives and routines. It is suggested that the way people are eating may affect the populations" health negatively. (Bublitz, 2010) revealed "that increases in the number of restaurant meals eaten resulted in a rise in weight gain". The more people eat out, the lower their quality of diet becomes and the higher the impact on their health (Morrison, 2011).

In South Africa 61% of adults are considered overweight or obese, according to The Star (The Star, 2012). "Excessive eating and unhealthy food choices are at the root of one of the most pressing health concerns facing the United States and much of the developing world (Bublitz, 2010). Health professionals hope to see that if this information is provided towards customers, will inform them of the nutritional contents of specific foods, as it would help them choose healthier, more nutritious meals and assist in their diets (Morrison, 2011).

By introducing nutritional information it would benefit individuals who eat out, as improving the dietary habits is an important step in improving public health (Josiam and Foster, 2009:876-891). From previous studies, we know that people underestimate the fat contents of menu items. One study has shown that the impact of state menu labelling law on population weight gain in LA, California, USA, says "there is an increase consumption of calorie dense foods outside the home" (Kuo, Jarosz, Simon & Fielding, 2009).

Results confirm that a relationship does exist and it is important to understand the link of eating out, weight gain and health concerns. South Africa is not alone in this „global obesity trend". If individuals pay more attention to what restaurants offer, they have the ability to better their health and benefit positively when dining out.

2.2 Nutritional Knowledge and the lack thereof

"In an increasingly obesogenic environment, nutritional advocates and public health organisations are urging consumers to be more aware of their eating decisions, encouraging moderation and a focus on healthy eating" (Bublitz, Peracchio & Block, 2010:9-258). Media has emphasized growing obesity, and with this the public are gaining knowledge and information about food decisions, thus being more diet conscious (Bublitz, Peracchio & Block, 2010; 3).

In South Africa the demand for NI is not as high as it should be, this could be due to the fact that people are unaware of nutritional information and lack health knowledge to support their daily diets. The dilemma is that most people lack the knowledge to grasp certain facts regarding nutrition, and food consumption therefore NI is of importance, and vital for consumers to understand (Heike and Taylor, 2012:151). It is important to get as many people familiar with engaging healthy eating behaviours as it is for these exact reasons that healthy positive food decision making should be promoted.

Due to poor eating habits, consumers are showing that they are unable to choose, and do not have the discipline to allow themselves these behaviours (Thomas Jr., 2006). They are seen as having no self-control (Bublitz, 2010). Many consumers may need a combination of general nutrition education jointed with NI presented in restaurants.

Results have shown that by providing this information in restaurants it would enhance customers" confidence levels and improve their images of the information"s source.

2.3 Nutritional Labelling/Menu Labelling

„Menu labelling is said to reduce customers" intentions to purchase food items that are high in calories and fat, especially when there is a larger difference expected between the perceived and actual food contents" (Larson, 2009). In South Africa due to the populations weight gain, NI has become an open topic to be incorporated within all restaurants, not only chains or franchises.

Menu labelling was introduced in attempts to help customers make informed decisions on foods and reduce the effects of obesity and weight gain. We already know that descriptive wording of menu items also affects choice and can increase sales by more than 10% (McCall and Lynn, 2008:439-445). By providing nutritional information in an easy to follow format and clearly distinctive pattern it would assist people in choosing healthier and more nutritious dishes (Morrison, 2011).

Roberto (2009) some consumers support the NI labelling and nearly half report it would influence their food choices. 64% of participants felt that restaurants should offer this information while 19% said they should not and 17% were not concerned.

Simply posting NI does not always get consumers to make healthier choices, so researchers tried colour coding. In a study by Lusk (2006) experiments were done based on three menu variations. One was controlled with no changes, one showed the caloric amounts only and the last showed calories as well as incorporated the traffic light symbols. These symbols showed what was healthier in green, less healthy in orange and least healthy in red. It incorporated price alterations also which showed what people eating out were likely and willing to spend, thus looking at the value for money and the industry sales. It was deduced that more people wanted to see the traffic light symbols as well as calories instead of just the calories as it gave them a sense of willingness to pay for what they know. The traffic light symbols are simple and do not have any numeracy demands, thus gives people insight into what not to eat. These had more of a profound effect on customers" choices and led to choosing meals with lower calories. Even though the changes were small they were still visible, so with that in mind if everyone started to take notice of these symbols and information it would benefit them and hopefully increase their opinions on their food choices.

In a NY University survey where 821 adult responses in a low income restaurant were studied, 27% of customers who saw this labelling said it did influence their choices while 88% said they in fact consumed fewer calories in response to these labels (Morrison, 2011). This therefore shows us that when people are given the correct nutritional knowledge they can apply it to their own diets.

In the Research Synthesis (2009) "focus group research of adult restaurants customers showed that most consumers are interested in having NI available even if they would only use it occasionally" (Larson, 2009).

Nutritional labelling is not compulsory for food items within full service restaurants, this may be why certain people choose what they do when dining out, and this may also be due to a lack of nutritional information and health knowledge. "Labels can help people sometimes if they have the knowledge or motivation to use the information, which may or may not be in a format they can understand" (Rotfield, 2009:375) (Heike and Taylor, 2012). This quote suggests that this issue is full of complexities and is of importance for consumers" education and protection. That is why introducing calorie labelling at the point of purchase is viewed as the newest public health method to decrease obesity and weight gain (Levy, 2012)

Calorie labelling may assist frequent diners in making healthier food decisions, or just help them in realising what is better for them to consume. Restaurants should therefore include specific facts informing consumers about the total calorie intake they should be allowed, whether they choose to use this information is up to them but hopefully it will make them more aware of what food items are out there (Morrison, 2011).

Getting individuals to become familiar with NI and subsequent dietary values is important. In relation to using NI to influence ones food choices, it seems that people like simplified and easy to follow nutritional labels (Heike and Taylor, 2012). These findings were supported by other studies concerned with preferences of NI and was indicated that the more information offered was better, but other studies show that too much information resulted in poorer performance due to the ease of using this information to make food choices (Heike and Taylor, 2012).

Holdsworth M 1998 & Steenhuis I 2004 (Holdsworth, 1998) agree that "labelling healthier items within the foodservice sectors may influence behaviour, promoting healthier dieting patterns" (Vyth, 2011). If consumers have no intentions of eating healthier food items or altering their diets and intakes they will not pay attention to the use of NI and labelling, this was visible from the results which suggested that only health conscious consumers would usually take note of this information (Vyth, 2011).

Claude Fischler, sociologist with the French National Research Centre for Scientific Research (2004) stated "If you are what you eat and you don"t know what you"re eating, do you know who you are?" (Heike and Taylor, 2012).

In this study it looked at finding out what customers want to see most on menus.

2.4 Restaurant Preferences versus Customer Preferences

An important question is; are people eating out to live, or do they eat out for pleasure? Most people live in rich food environments where there is a variety of foods available so it is seen as they eat out for enjoyment (Bublitz, 2010).

"Customers expect restaurateurs to provide information on the calories, sugar, protein, fat, and carbohydrate content on menus", (Mills and Clay, 2001:69-82); (Mohd Zahari, Din & Shariff, 2010:333 - 346). According to (Hwang and Lorenzen, 2008:270-276) providing nutritional information on restaurant menus aids the consumers in making healthier choices and they were in fact willing to pay more for food they perceived as healthier.

Emotion and moods may interrupt people"s consumption decisions, and these can vary during the day. It requires consumers to balance their emotions which is challenging for many. Positive emotions can impact on health and may make customers more focused on the NI attributes & features of food, while with an increased variety of food items, there is usually a strong direct increase of how much people eat, thus resulting in over eating. Negative emotions like stress can also lead to over eating, while for others it may lead to eating much less (Bublitz, 2010). The more knowledge people have about how their emotions play a part in their eating decisions, the more they will be able to manage their own diets (Bublitz, 2010).

Personal factors including motivation and the ability to recognise the health concerns and connections are a significant part of the consumers" choice. Nutrition information is affected by personal factors such as "one"s ability, motivation and knowledge levels" (Heike and Taylor, 2012). These are seen as vital factors in accessing the involvement and nutrition knowledge when processing this information. They also found that "consumers may only read and process this kind of information if it applies to their personal beliefs" (Heike and Taylor, 2012). These results were consistent with previous research that showed that changes in nutritional labelling formats improved peoples abilities to process this information. This in turn helps individuals to understand the health risks involved and weigh up their food decisions. People are said to eat out more due to social and environmental factors, which is a concern for people"s health, as they eating for reasons other than hunger (Bublitz, 2010). Social factors may attract or repel individuals to eat certain food items especially when one sees another enjoying their meal. This could promote an individual"s enjoyment and alter their decisions regarding the food and experience they have all together Socially, some families place influence on the younger ages to shape their dietary habits and mould individual taste preferences, while friends, business work environments and different occupations also a play a part in peoples eating tendencies (Bublitz, 2010). Gender "Coll, Meyer, and Stunkard; agreed that men purchase more than women. It was also identified that when men ate with other men they ate the most and when women ate with me they ate the least" (Brindal, 2010). Bublitz (2010) supports this due to evidence findings that women are more concerned with their weight status than men. "Women are more likely to use NI than men" (Larson, 2009).

If customers view the restaurant menus as healthier and more valuable they "may be willing to pay more for that increased value" (Hwang, 2008). If this occurs restaurants can attract customers successfully and gain sound revenues, therefore achieve both financial and nutrition goals (Hwang, 2008), however if food is perceived as unhealthy customers will not be willing to pay more. Hwang (2008) confirmed past studies stating that the more NI shown, the more preferred the restaurant was to customers. In his study "81% of participants said they do read NI when they shop, 93% felt that restaurants should include NI and 90% said they would make use of it" if it was placed in restaurants.

Overall we can therefore conclude that NI can benefit both customers and restaurateurs, as customers gain confidence as well as improve on how they view the restaurant (Hwang, 2008).

From the current literature review it has shown that there is a significant change in the western eating behaviours, as the purchase and consumption of restaurant foods has increased. This paper shows the Hospitality and restaurant industry efficient and effective ways and ideas of how to incorporate NI in order to assist consumers making better life changing choices. It shows how restaurants and customers perspectives differ regarding menus and the detailed information that is viewed as important. It focused on how these factors could influence restaurant sales and overall productivity, therefore satisfying both the customers and restaurateurs for future improvement, growth and development, as well as promoting health awareness amongst the general population.

3. METHODOLOGY

3.1 Introduction

This study was intended to use both Quantitative and Qualitative approaches to assess whether or not nutritional information was a necessity within Johannesburg restaurants.

To determine and learn more about the effects and customers attitudes on nutritional information (NI) in restaurants the researcher conducted a three week study using two restaurants. Five menu items were chosen from restaurant A, and eight menu items from restaurant B, in order to answer these four questions: (sub research questions).

- Does having visually attractive pictures influence consumers food/meal choices?
- Does having NI, health icons, allergies etc, in restaurants influence consumers" food/meal choices?
- Does having NI in restaurants increase the sales and productivity of the restaurant?
- Does the use of NI make customers more aware of their health and diets? And therefore improve their nutritional knowledge?

In the questionnaires given out, demographic questions were included, as they gave good indicators of food choice behaviours. These included; age, gender, ethnicity, highest educational qualification, how often customers ate out etc. Participants were also asked how many members of the family were dining out together at that specific time and whether or not the provision of this information would influence their food choices.

3.2 Research Methodology

Restaurants A and B menus contained similar dishes of breakfast and lunches, so some dishes were chosen from breakfast and others were chosen from lunch. These dishes were then represented in the form of an attractive picture in the 2nd week, and in the 3rd week nutritional information such as health icons, allergies and so forth were placed additional to the food item descriptions.

Both restaurants are middle to upper class, with a variety of customers, differing in age, gender, and racial status. Neither of these restaurants had any form of visual displays, icons or nutritional information on their menus, thus they were ideal to use in this investigation.

For the first week the restaurant menus had no changes or additions made to them, in order to see what sold the most and what did not sell as much, therefore what was less popular. From these findings, as stated above each of the restaurants chose between 4-8 menu items to modify within the second and third weeks. The dishes that were chosen for weeks 2 and 3 were displayed using attractive pictures of these meals on a separate laminated leaflet accompanied by the original menus. The 3rd and final week the addition of NI, and health icons (allergies etc), were displayed on these leaflets additional to the pictures, to see the differences from the 1st and 2nd weeks, and from this a comparison over all 3 weeks. By doing this and reviewing the patterns/changes if any, it showed the restaurant owners and managers the customers" attitudes and behaviours behind their eating choices, and how their psychological knowledge and insight on health affects the options they choose.

All data was structured through the use of mixed methods; (qualitative) face to face interviews, document reviews, individual concealed observations and (quantitative) administering written questionnaires/ surveys. These were used to generate a table of the before and after comparisons, of the differences throughout the three week study

Research/Sample Design

Two types of sample groups were examined; purposive sampling and convenience sampling

The Population of the study was divided into two groups:

1) A convenience sample was utilized and focused on all the customers who ate at these restaurants, as the selection process was aimed at the restaurant customers specifically and not focusing on the general public. Every customer at the restaurants was given a questionnaire, and it was up to them if they wished to fill it in.

2) A purposive sample used to choose the two restaurants. Interviews were conducted with the restaurants management personnel to ensure the questionnaires answered were reliable, relevant and informative.

Quantitative

Questionnaires were created for, and handed out to all customers at both the restaurants.

The quantitative approach (questionnaires for customers at the two restaurants) helped to understand the views, behaviours and attitudes of what foods people should choose while dining out. The distribution of questionnaires was aimed at customers who ate at these restaurants only. The questionnaire was created and used to capture the data. Permission from the restaurant managers and owners was attained, and names of participants were treated privately and confidentially. These customers were given the surveys as the menus were handed out and they could fill them in when and if they wished. Each participant received a one page A5 back to back booklet questionnaire. This included the cover page which defined the purpose of the study. The initial sample consisted of 60 participants per restaurant and therefore a total number of 180 questionnaires from each restaurant were completed over the three week study, giving a final amount of 360 questionnaires all together.

Derived from the results of the questionnaires, completed by all the customers determined their opinions and points of views regarding NI and pictures when being incorporated in JHB restaurants and on their menus. Feedback included their dining experience and how the NI, health icons and pictures influenced their meal choices. The customers" who were given these questionnaires constituted a "convenience sample" This allows the researcher to save money and time as the respondents are already available for the researcher to approach them. The customers who did complete the questionnaires were only found within these two restaurants, and this made the study more efficient and economical.

Qualitative

Interviews held with management personnel from both restaurants:

The qualitative approach (interviews held with management personnel) helped to understand what made these two restaurants successful, if customers" choices would be influenced and to see in fact if their sales would positively change due to what customers select from the menus

These were conducted in person with the restaurant managers/owners a week after the completion the three week study. Individual statements were obtained when the number of people or the timing was impractical. The interviews were recorded and transcripts were written out based on the same questionnaire format.

These interviews asked why the management saw the study as useful, how they thought it would assist their overall business as well as what they have seen, identified, and have realised about the integration of nutritional information within their restaurants.

The general ambience, food presentation and service was noted by the researcher during field work and observed from the staff and management as well. Managers from the two restaurants made up the purposive sample. The restaurant & managers were specifically selected as a result of their qualities and knowledge regarding the way they chose to incorporate NI and appealing pictures within their restaurants.

The data collected from these interviews revealed certain reasons for wanting to introduce NI within restaurants, both on a personal level, as the managers may have a desire to promote this information onto their customers, as well as on the economical side of increasing and enhancing clientele and promoting the restaurant business as a whole.

Collecting the Data

The following steps were taken for collecting the data:

Step1. Consent was given to interview the managers and owners at the two designated restaurants.

Step2. Details and specific information on the ways in which both these restaurants promote and sell their foods to customers, as well as their sales and productivity was covered.

Step3. The researcher ate at both the restaurants to experience „their dining out concepts". Written notes pertaining to the taste, presentation, service, price etc of the chosen dishes were identified and noted.

Step4. Questionnaires were handed out to all customers who ate at the restaurants to get their feedback on how these factors influenced their choices and to find out if there was in fact a demand for these factors. Interviews were held with the two restaurants managers and owners, therefore one from each restaurant.

Step5. Research from books, journals, articles and internet websites added to the information of this extensively interesting topic and was accumulated for a comprehensive literature review

Analysing the Data

Step1. Field notes were gathered as well as photographs were taken of the two chosen restaurants & their specific food items. Questionnaires for customers and interview write ups with owners/managers were organised.

Step2. The information was then separated within the two restaurants which divided into specific sub questions which owners were to answer.

Step3. Data was then assessed with each restaurant and answers were based on these questions that were asked.

Step4. Research from the sub questions were created for a well-structured research study.

Step5. The main concepts of each restaurants question were discussed in depth, and this data was further manipulated and broken down

For organising and interpreting the quantitative data, Microsoft word was used to place the data from the surveys onto computer software, in order to distinguish the before and after comparisons and to plot the data on graphs. Microsoft Excel, a spreadsheet programme was used to draw up the questionnaires and the use of SPSS software which was incorporated to assist with the statistical side of the data, and thus a consultation was held with STATCON.

SPSS is a system which focuses on statistical packages. This was helpful as it analyzed data, making the coding and interpretation of results easy to understand. It incorporated the Chi-square test "which is used to explore the relationship between two categorical variables." "It compares the observed frequencies or proportions that occur in each category, and is based on a cross tabulation table, with the cases classified according to the categories in each variable" (Pallant; 2007:214) (Pallant, J. 2010).

The questions consisted of mainly close-ended questions and one open-ended question, where customers were able to give feedback if they wished to. Descriptive statistics were tabulated in an organised fashion so all values were formatted clearly to provide objective answers. Additional charts and graphs were drawn to identify the differences from both restaurants as well as from week to week.

Organizing and interpreting the qualitative approach and data, Microsoft Word 2007 was used to write up the questions for both restaurant interviews. These interviews with the managers/owners were recorded and transcripts were captured by the researcher and written out according to the outcomes seen over the three weeks. The interview analysis, was conducted with both restaurant owners, and was divided into several questions in order to maintain a consistent sequence for all questions asked. The results from these described how they established the way they chose to integrate the use of the NI, health icons and pictures within their restaurants. Both interviews were compared with the questionnaire results, to see if there were any differences. The researcher also performed observational research to identify any changes seen personally.

Ethical Considerations

The purpose of the study was clearly shown to all respondents before any comments and results were taken. In order to keep it private and confidential all participants were kept anonymous when filling out the questionnaires, so no names or personal identity was needed. These people were treated with respect and dignity.

All responders and the managers of the two restaurants were given a cover letter detailing out what the research entailed, and outlined what this study was set out to do. The management personnel of the two restaurants were approached by the researcher and consent to go ahead with the interviews was given. The purpose of the study was communicated to the owners/managers and this prevented unnecessary misunderstandings and errors from occurring. All who took part in the study were thanked and acknowledged by the researcher for their contribution they brought forward to the study.

This study followed all the rules and regulations of the University of Johannesburg, and the Universities letter head was displayed on all the questionnaires.

Field Work

Field work was completed by the researcher by viewing, studying and researching the menus and the backgrounds of locally privately owned restaurants. After finding the two suitable restaurants, the owners/managers were contacted by telephone calls to give consent and permission to use their restaurants as part of the study. Shortly after this, the researcher made appointments to meet with the management personnel, speak to them personally and conduct the necessary interviews which conveyed what the study was all about. Once the approval from the managers/owners had been given, questionnaires were printed and handed out to all customers who ate at these restaurants, and results of this feedback were recorded. These questionnaires were; one for the 1st week when no changes were done, this questionnaire focused only on what customers would consider, while for weeks 2 and 3, they focused on what the customers were actually looking at ordering/choosing from the menus

During the observations done by the researcher, notes and transcriptions were written in order to provide additional valuable data which contributed to the topic on hand. The researcher noticed that with some people the visual pictures on the leaflets for specific meals made a huge impact on customers" intention to order those meals. Some of these people even changed their minds on what orders they wanted, showing that it altered their behaviour behind their food choices. The researcher can say that through the use of these pictures, it did in fact impose upon customers as it caught their attention.

4. RESULTS

4.1 Quantitative Analysis-Demographics Results

Statistics are an average across both restaurants.

- 60.34% Female
- 74.5% Middle Aged (ages 25 – 54)

- 83.27% White • 80.6% Highly educated
- 39.2% Majority ate out more than 6 times a month

It can be noted that from the feedback and high performance of females it could be a result of the venues chosen, or because females are said to be generally more nutritionally conscious of what they eat as opposed to men. The same was identified according to Larson (2009:3) study that mentioned how "women are more likely to use nutritional information (NI) than men" & according to Dubbert (1984:86) study it found the same. This needs to be taken into consideration when looking at how one answered the questionnaires.

The age groups were a representation of who was present at the exact time the study was carried out. These results were consistent with other research done, such as (Josiam; 2009:882-883) study. Race showed a continuous higher percentage of customers eating at these restaurants over all 3 weeks were white. Individuals who ate out here were highly educated and thus logically following this, they should be knowledgeable and more consciously aware of what is right and wrong to consume. This was reinforced by (Josiam; 2009;883) & (Heike & Taylor; 2012:122-123). Most individuals were seen to eat out more than 6 times a month at both restaurants. This could be due to the high amount of working people who may drop in for a quick meal or on the go coffees and meetings.

Due to a person having a higher educational qualification and better employment positions, with that comes higher incomes and the ability to afford dining out at restaurants just like these 2 chosen, as well as affordability to take families or others out to dine with them. This was proof from (Heike & Taylor; 2012:140-142) as they support that the higher the income the more likely individuals were to use and compare NI.

In conclusion, from the statistics of demographics, that the racial group was predominantly white females. This means that different responses may arise when viewing other races, and possibly males specifically.

4.2 Findings from Questionnaires

From the above figures 4.1 a & b which looked at how pictures influenced customers" food decisions, a positive change was seen from week to week, as having the actual pictures in front of customers made more of an impact on their opinions.

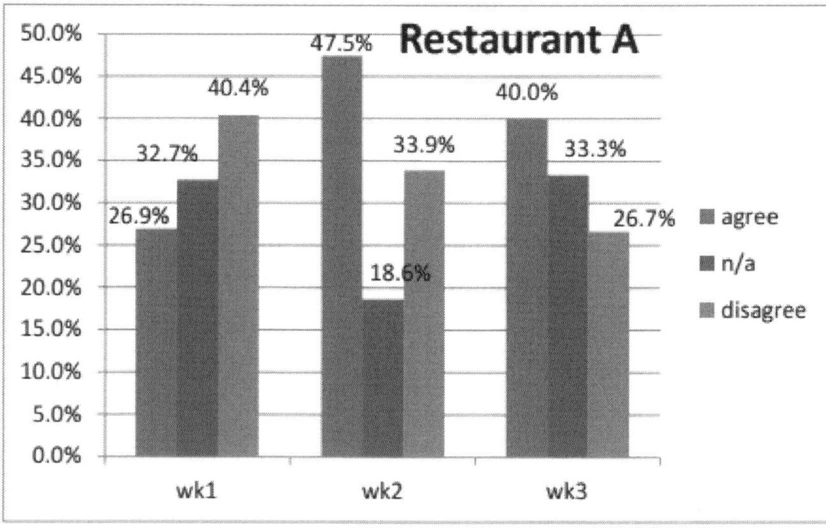

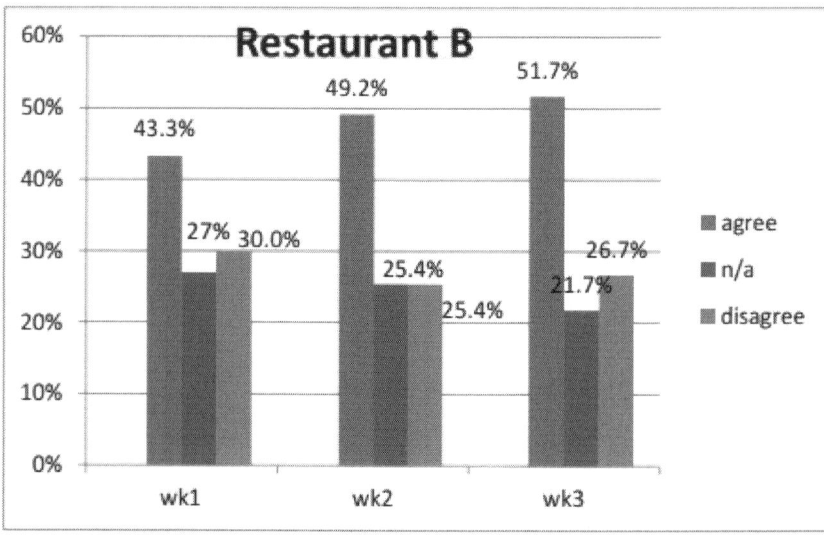

Question 7.1. Visually attractive pictures on the menu influence my food choices.

Table 1.1 and Table 1.2 shown below are the cross tabulations that were analysed. Table 1.1 and Table 1.2 was a cross tabulation of question 8 and 7.1 from the questionnaires. Table 1.3 and Table 1.4, was another cross tabulation that used question 8 and 7.2 from the questionnaires. Both these were tabulated with the „neither agree nor disagree" option being eliminated and therefore only „agree" and „disagree" options were given. Table 1.1 and Table 1.2 illustrate what percentage of respondents agreed or disagreed with question 8, and what the percentage was of these respondents who chose their meal choice from either the main menu or the separate leaflet which displayed the visually attractive pictures. Table 1.3 and Table 1.4 illustrate what percentage of respondents agreed or disagreed with question 8, and what the percentage was of these respondents who chose their meal choice from either the main menu or the separate leaflet which displayed the nutritional information and health icons.

Table 1.1 Cross tabs of question 8 & 7.1 Restaurant A

Restaurant A		**Q:8**		
Q:7.1		Leaflet %	Main Menu%	Total%
	Agree	39.3	60.7	100
	Disagree	15	85	100
	Total	29.2	70.8	100

Table 1.2 Cross tabs of question 8 & 7.1 Restaurant B

Restaurant B		**Q: 8**		
Q:7.1		Leaflet%	Main Menu%	Total%
	Agree	28.6	71.4	100
	Disagree	13.3	86.7	100
	Total	23.3	76.7	100

From Table 1.1 and Table 1.2, data that was captured giving the Chi-square test showed a continuity correction at restaurant A of 2.259 and sig value of .133, and at restaurant B .560 & .454 respectively. This was seen as insignificant as the sig value needs to be 0.5 or smaller to be of significance. The Phi values were 3.68 and .244 which placed it in having a small to medium effect. "As from 0.10 it is viewed as a small effect, 0.30 is a medium effect and 0.50 is seen as a large effect" (Pallant, 2007:217) (Pallant, J. 2010). From this it can be further concluded that the minimum expected counts were 5.83 and 3.49 respectively, and needs to be 5 or more, so from this cross tabulation in one regard it was identified as significant and valuable but not enough proof/evidence was obtained for the general public to use this as an evident factor, and therefore overall was not valid enough for continued use.

Figure 4.2 a & b looked at how the NI influenced their choices, and a negative influential effect was seen with the inclusion of NI over the 3 weeks. It seemed to turn people off and therefore pushed them away from considering this information, as they seemed to not be aware of the caloric amounts within specific food items. More positive results such were seen at restaurant B which are more in line with other studies by (Vyth; 2011:132) & (Roberto; 2009:5).

Data that was captured for Table 1.3 and Table 1.4, using the Chi-square test showed a continuity correction at restaurant A was .085 and sig value of .770 and at restaurant B 1.185 and .276 respectively, this was seen as insignificant for the same reason as mentioned above. From this it can be concluded that the minimum expected count was 3.16 & 1.50 and needs to be 5 or more, so from this cross tabulation not nearly enough proof/evidence was obtained and therefore was not significant either.

People care about nutritional information, as there was a connection seen between health concerns and what people eat.

This study founded similar findings to Morrison (2011:10-12) & Heike and Taylor (2012:120-121) which helped to prove that there was a connection between eating out and health concerns, shown in Figure 4.3 (question 7.4). Restaurant A, due to the lack of nutritional knowledge fewer people agreed to this which was a bad reflection on the respondents. Restaurant B most agreed and therefore we can conclude that these respondents were more knowledgeable, or perhaps restaurant A respondents were just not as interested.

Question 7.2. The use of health icons on menus influences my food choices

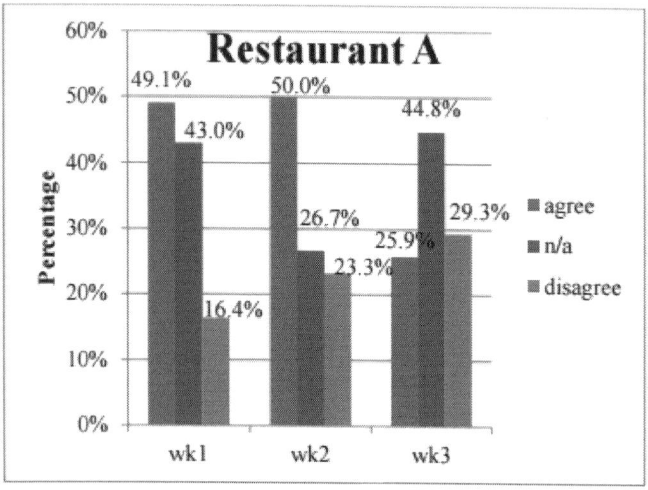

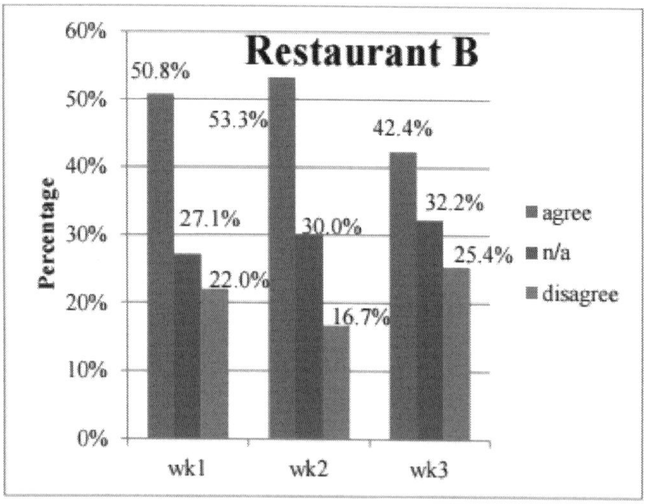

From Figure 4.4 a & b, (question 10) displayed factors that had to be arranged from most important to least importance. It was seen that taste was the top priority, followed by food quality, then service, then food presentation, then health and lastly price.

Table 1.3 Cross tabs of question 8 & 7.2 Restaurant A

Restaurant A		Q:8		
Q:7.2		Leaflet%	Main Menu%	Total%
	Agree	28.6	71.4	100
	Disagree	17.6	82.4	100
	Total	22.6	77.4	100

Table 1.4 Cross tabs of question 8 & 7.2 Restaurant B

Restaurant B		Q:8		
Q:7.2		Leaflet%	Main Menu%	Total%
	Agree	16	84	100
	Disagree	0	100	100
	Total	10	90	100

Visual aids and/or NI do not impact on restaurant sales.

Figure 4.5 (question 7.5) depicted what percentage of people would be willing to pay more for a fresher, healthier meal. In studies by (Hwang; 2008:271) it was found that people would be willing to pay more if perceived as more valuable. Conversely, others who did not see the quality of food in these ways were not willing to pay more. The latter was not as evident here as most people at both the restaurants agreed to this, saying that they would be willing to pay more for better quality food items.

In conclusion to all the above, most people were interested and captivated by this information, as it offers reasonable insight into showing them the right directions to make better food choices. However others do not see it as a requirement at all. It definitely needs more complex research to be investigated, better statistical results and stricter health policy assistance in order to make a real impact on the public"s views and opinions regarding this topic

Question 7.4. I believe there is a connection between health concerns & what people consume while eating out.

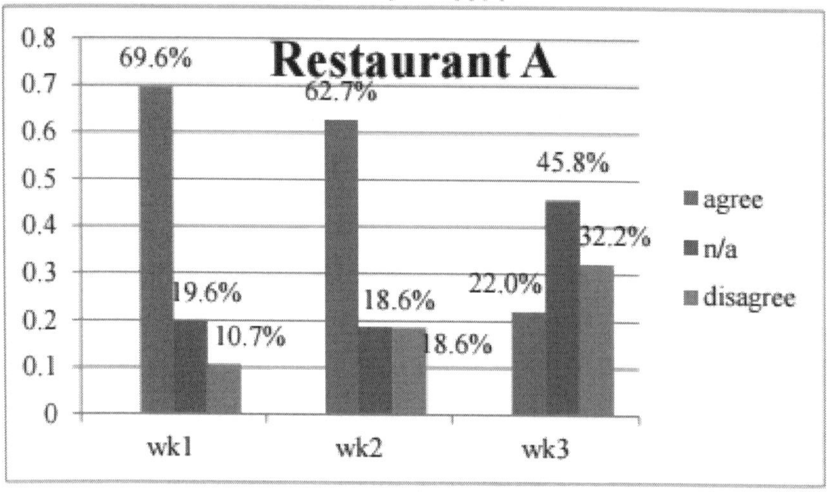

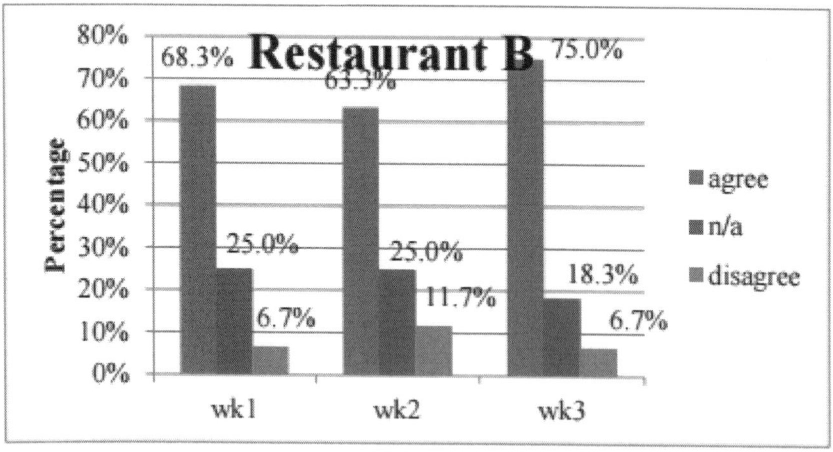

Question 10.1- How important is health

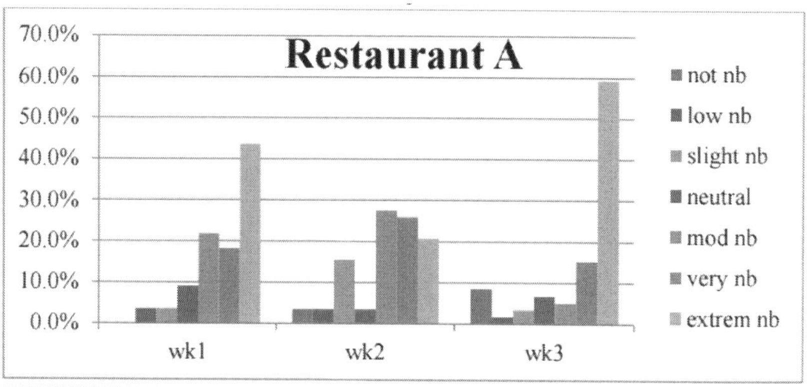

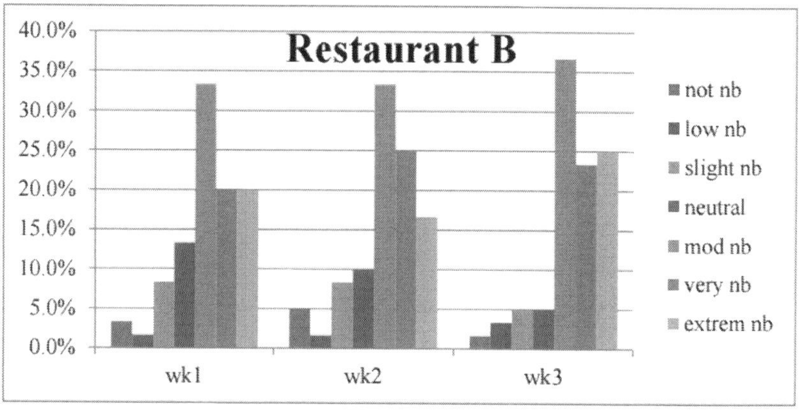

4.3 Qualitative Analysis- Discussion of Interviews

The results from the interviews suggested that restaurant A would use the pictures and health promoting icons but not caloric amounts, and restaurant B would use this information to promote people"s health and nutritional knowledge.

Restaurants A and B did not feel they were the best choice of restaurant for such a study, yet both felt similar to consumers in that it is important to know and be aware of such information, but not with such a demand as one should know what they should and should not consume already.

Lastly it was seen that overall the pictures made more of an impact to both restaurants due to nutritional information not being well understood.

Question 7.5.I am willing to pay more for a healthier, fresher meal.

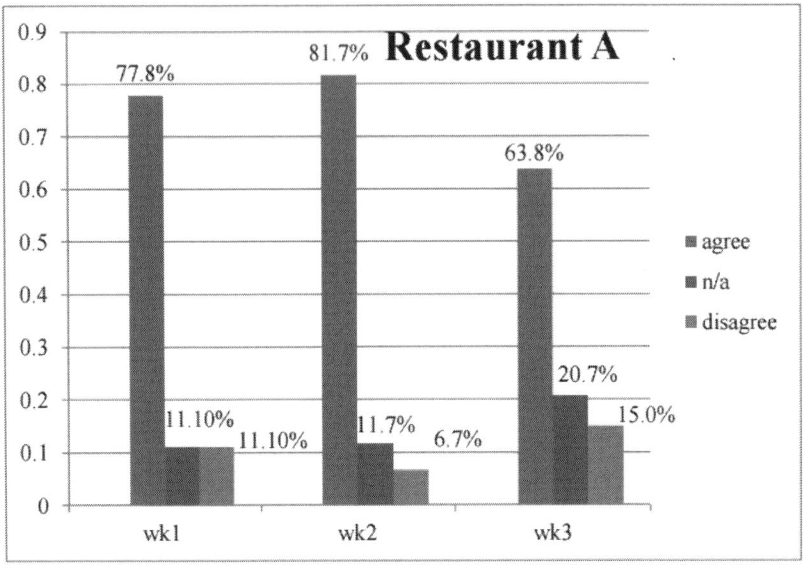

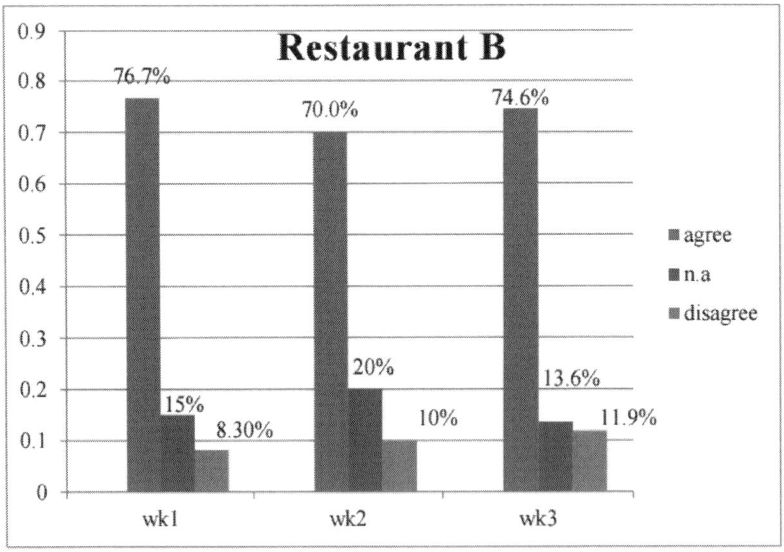

Therefore incorporating information on menus is vital for customers to make better choices. These findings were in line with Panitz (2002:3) & McCall (2008:440) studies as both believed that menus should be graphically fun yet functional at the same time, as they show customers how they should perceive the restaurant and the foods they wish to order.

This can be supported by Heike & Taylors study (2012), where overall nutritional knowledge is an important factor in order for customers to make reasonable food choices & benefit their health in the future.

4.4 Comparison of interview and questionnaire results

In comparison as noted above, on average out of the sample most people thought it would influence them but over the weeks this response was seen to change as many did not know whether to agree or disagree with some of the statements. This showed a lack of nutritional knowledge and health concerns for individuals overall

This study was purely a generalisation of what customers would choose or not choose due to what they see when dining out. To conclude further research would be a good idea and needs to be undertaken to sample a larger group over a longer time period in order to get a wider spectrum of results.

5. CONCLUSIONS AND RECOMMENDATIONS

5.1 Conclusions

In conclusion, the methodology and processes used to bring all the relevant data to this study, made a clear understanding as to the ways the researcher went about collecting and analyzing the data.

Comparing the results of both restaurants to previous studies done, one can draw up the conclusion concerning whether Nutritional Information, health icons and pictures are actually needed within restaurants for the assistance of customers" food choices, the improvement of their diets and future lives, as well as increasing the sales and productivity levels of restaurants. This then benefits and satisfies both customers and restaurateurs as a whole.

Consumers were seen to want to have this information available-demand (might not use it but if they choose to it"s there). Pictures were viewed as having more of an impact over Nutritional Information and there was a definite lack of nutritional knowledge, suggesting individuals are uneducated about Nutritional Information, this further leads to the possibility of scaring off consumers.

5.2 Recommendations

For this study it is recommended that further research/experiments should be done using more restaurants and having a broader variety of menu items. Increasing the sample size would be another option as well as using the same sample from start to end.

It is also recommended that longer time frames and different factors are used, such as the general public, not middle-upper class only.

Lastly restaurants can help by placing promoting nutritional facts (daily intakes etc) to guide customers to choose better food items and assist individuals to start learning about Nutritional Information.

REFERENCES

1. Bublitz, M.G., Peracchio, L.A. & Block, L.G. (2010). Why did I eat that? perspectives on food decision making and dietary restraint. Journal of consumer psychology, 20(23):9-258.

2. Dubbert, P.M., Johnson, W.G., Schlundt, D.G. & Montague, N.W. (1984). The influence of caloric information on cafeteria food choices. Journal of applied behavior analysis, 17(1):85.

3. Food and Drug Administration. (2012). Available from: http://www.fda.gov/Food/LabellingNutrition/LabelClaims/ucm187320.htm#front. (Accessed September 20).

4. Food and Drug Administration. (2012). Available from: http://www.fda.gov/ora/inspect_ref/igs/nleatxt.html. (Accessed

September 12). Heike, S. & Taylor, C.R. (2012). A Citical Review of the Literature on Nutritional Labelling. 2012, September 10.

5. Holdsworth, M. & Haslam, C. (1998). A review of point-of-choice nutrition labelling schemes in the workplace, public eating places and universities. Journal of human nutrition and dietetics, 11(5):423-445.

6. Hwang, J. & Lorenzen, C.L. (2008). Effective nutrition labeling of restaurant menu and pricing of healthy menu. Journal of food service, 19(5):270-276.

7. Jones, C.S. (2009). Taking up space? how customers react to health information and health icons on restaurant menus. Journal of food service business research, 12(4):344-363.

8. Josiam, B. & Foster, C. (2009). Nutritional information on restaurant menus: Who cares and why restauranteurs should bother. International journal of contemporary hospitality management, 21(7):876-891.

9. Kuo, T., Jarosz, C.J., Simon, P. & Fielding, J.E. (2009). Menu labeling as a potential strategy for combating the obesity epidemic: A health impact assessment. Journal information, 99(9):. Larson, N. & Story, M. (2009). Menu labeling: Does providing nutrition information at the point of purchase affect consumer behavior. A research synthesis.healthy eating research national program.robert wood johnson foundation, .

10. Levy, D.E., Riis, J., Sonnenberg, L.M., Barraclough, S.J. & Thorndike, A.N. (2012). Food choices of minority and low-income employees: A cafeteria intervention. American journal of preventive medicine, 43(3):240-248.

11. Lusk,J.L. Effects of 'fat tax' & calorie information on restaurant food choices. 2006, September 5.

12. Magoulas, C. (2009). How color affects food choices.

13. McCall, M. & Lynn, A. (2008). The effects of restaurant menu item descriptions on perceptions of quality, price, and purchase intention. Journal of foodservice business research, 11(4):439-445.

14. Mills, J.E. & Clay, J. (2001). The truth-in-menu-law and restaurant consumers. Foodservice research international, 13(2):69-82.

15. Mohd Zahari,MS., Din, N.,Shariff,SM (2010). Nutritional information in restaurant menu: What do malaysia customers think? Institute of interdisciplinary business research, 2(4):333 - 346.

16. Morrison, R.M., Mancino, L. & Variyam, J.N. (2011). Will calorie labeling in restaurants make a difference? Amber waves, .

17. Pallant,J. (2010). SPSS survival manual: A step by step guide to data analysis using SPSSOpen University Press.

18. Panitz, B. (2000). Does your menu attract or repel diners. Reading between the lines: The psychology of menu design.restaurants USA, 881-87.

19. Roberto, C.A., Agnew, H. & Brownell, K.D. (2009). An observational study of consumers' accessing of nutrition information in chain restaurants. American journal of public health, 99(5):820.

20. Roberto, C.A., Larsen, P.D., Agnew, H., Baik, J. & Brownell, K.D. (2010). Evaluating the impact of menu labeling on food choices and intake. American journal of public health, 100(2):312.

21. Swart, R., Sanders, D. & McLachlan, M. (2008). Nutrition: A primary health care perspective. South african health review, 129-148. .

22. The Star. (2012). Available from: http://www.iol.co.za/the-star/more-than-half-sa-adultsoverweight.html. (Accessed September 25).

23. Thomas Jr, L. & Mills, J.E. (2006). Consumer knowledge and expectations of restaurant menus and their governing legislation: A qualitative assessment. Journal of foodservice, 17(1):6-22.

24. Vyth, E.L., Steenhuis, I.H.M., Heymans, M.W., Roodenburg, A.J.C., Brug, J. & Seidell, J.C. (2011). Influence of placement of a nutrition logo on cafeteria menu items on lunchtime food choices at dutch work sites. Journal of the american dietetic association, 111(1):131-136.

6

A DETAILED COMPARISON OF FINLAND AND INDIA THROUGH HOFSTEDE & GLOBE STUDY

Sivasubramanian Narashima Boopathi

Doctoral Student, Department of Management,
University of Vaasa, Vaasa, Finland.

ABSTRACT

The aim of this study is investigate the cultural differences and similarities between India and Finland through the cultural dimensions proposed by Hofstede and GLOBE study. To understand the differences between business and organizational cultures of Finland and India cultural dimensions play a significant role in business success. Though there are various people have contributed for differentiating the culture, the cultural dimensions proposed by Geert Hofstede has been first of its kind which opened up the imagination to differentiate the national cultures. Apart from Hofstede's cultural dimensions GLOBE study has been one of the most recent studies on cultural dimensions and leadership effectiveness. The cultural dimensions are significant in finding out the similarities and differences between the cultural values, practices and leadership style of Finnish and Indian organizational culture. This study will contribute to the academic literatures devoted to cross-cultural management and International management. This paper will offer novel insight to cognize how Finland and India differ in their values, practices and preferred leadership styles through the cultural typologies proposed by Hofstede and GLOBE study. This study will also have number of practically applicable points for the business leaders to manage their business in between India and Finland.

KEYWORDS

Hofstede, GLOBE, Cultural dimensions, India, Finland

1. INTRODUCTION

Business cultures differ from country to country due to the influence of its national culture. The national culture of a country is shaped by various factor rights from the origin of that country to the present day social changes. The cultural values of nations have significant influence in the day today life of people and as well in its social, economic, political and business environment. The impact of national culture on the management and organizational behavior of an organization operating in a particular country is unavoidable (Francesco & Gold, 1998:18). The organizations operating in an international business environment may have set of standard policies and rules but it is obvious that those policies are followed differently within its own organization as the people from different countries do things in different ways. Moreover, people within the organizations will also have different attitudes and behaviors due to the influence of their national culture (Trompenaars & Turner, 1998:7). For example people experience phenomena such as authority, bureaucracy, creativity, good fellowship, verification and accountability in different ways in different parts of the world due to the influence of their national culture (Trompenaars & Turner, 1998:3). Therefore, it is inevitable that national culture affects its business environment and business organizations as business must interface with people, either as customers, employees, suppliers or stakeholders from different parts of the world (Jones, 2007:2). Besides, due to influence of culture the leadership style and management methods also varies as there is interdependent relationship between leadership styles and cultural underpinnings that cannot be underestimated (Jogulu, 2010:706; Taleghani, Salmani & Taatian, 2010:92). In order to understand the differences between business and organizational cultures of different nation's cultural dimensions play a significant role. Hofstede (2010:31) defines "dimension as an aspect of culture that can be measured relative to other cultures. Cultural dimensions enable the provide ways of understanding behavior encountered in business situations that at first may appear odd, mysterious or inscrutable". Therefore, cultural dimensions play a dominant role in understanding the organizational culture of different nations.

In the above mentioned background this study will compare the cultural dimensions of India and Finland. The reason for comparing India and Finland is that India and Finland are both different from each other in its culture, history and traditions. In the recent years there is an increase in economic tie between India and Finland. Finnish companies are expanding their businesses rapidly in India to serve India's domestic market (Numminen, 2007; Finpro, 2007). At the same time the Indian companies are also expanding their operations in Finland. Comparison between India and Finland itself can lead to confrontations as there is huge dissimilarity between various aspects of Finland and India. India and Finland have specific cultures which intend the people from both of these countries to behave differently. Finnish culture can be considered as homogenous when compared with the Indian one

There are several differences exist between business culture and leadership style between Finland and India. First, in the Finnish culture supporting and delegating leadership style is preferred more, whereas, in Indians prefer more of a directing and coaching style (Routamaa & Debnath, 2011:898). Second, the leadership in India is based on paternalism where the leaders expected to behave like a father in family and inspire the employees (Rangnekar, 2004). Whereas, the Finnish leaders are expected to be straightforward and to show egalitarianistic behavior. Third, when comparing Finland and India based on Hofstede's (1991) cultural dimensions India is characterized as masculine culture with high power distance; whereas, Finland is characterized as the feministic culture with low power distance. Fourth, Finland is characterized as individualistic culture while India is a collectivistic culture (Hofstede, 1991; Routamaa & Debnath, 2011). Fifth, due to the vast cultural differences between Finland and India (Finnish culture is considered to be more homogenous and Indian culture is more heterogenic and culturally pluralistic in nature) the very effective Finnish leadership style in Finland may be extremely ineffective style for India. (Kets de fries, 2001 in Lamsa, 2010:140).

In this kind of dissimilar cultural environment between India and Finland, the perceptions, attitude and behavior of Finnish leader in India may be different from the perceptions and behavior from the Indian followers and other stakeholders and vice versa. Therefore, it is significant to compare both Finnish and Indian cultures as it will facilitate the comparison between two different countries which are totally contrasting to each other. Furthermore, this study will also be a noteworthy contribution to the literatures devoted to cultural dimensions, comparison of cultural dimension between two countries (East vs. West), Hofstede's cultural dimensions, GLOBE study, leadership styles comparison and Indian and Finnish organizational cultures comparison. This study will offer contributions to the literatures mentioned above by explaining how the factors such as power relations, leadership style, gender egalitarianism, future orientation etc is handled in Finnish and Indian culture. Additionally, this study will also give an insight how the cultural dimensions can influence people's behavior in the Finnish and Indian organization. Besides, the study will also offer several managerial implications by providing insights to the business leaders and Individuals from India and Finland and as well as for those business leaders who would like to work in this countries. Therefore, it is important to understand the differences between Finland and India through cultural typologies proposed by Hofstede and GLOBE study.

Though there are various people have contributed for differentiating the culture, the cultural dimensions proposed by Geert Hofstede has been first of its kind which opened up the imagination to differentiate the national cultures. Apart from Hofstede's cultural dimensions GLOBE study has been one of the most recent studies on cultural dimensions and leadership effectiveness. GLOBE stands for Global Leadership and Organizational Behavioral Effectiveness. GLOBE was conceived by Robert J.House in 1991. GLOBE study published the cultural dimensions and leadership effectiveness of 62 countries. GLOBE conceptualized the culture of 62 countries in terms of nine cultural attributes that are referred as dimensions. The typologies proposed by Hofstede and GLOBE may not completely reveal all the aspects of Indian and Finnish culture totally as there are so many subcultures, languages, customs, traditions and other cultural factors are embedded in them. However, these dimensions enable us to understand both the cultures generally through the scores and ranking of the dimensions. Moreover, discussing the difference between Finnish and Indian culture through the cultural dimensions will enable us to predict the behavior of the people from both the cultures, it will explain the reasoning behind the actions of the people from both the cultures, it can help us to look for universalities between both the cultures, and it can create an understanding on standardized policies (Lewis, 2006:29, in Edinger, 2011:14). Moreover, understanding the cultural differences and similarities between Finland and India through Hofstede and GLOBE study will improve the bicultural interaction between the Indian and Finnish business leaders (Fougere, 2004:19). Besides, in-depth understanding of Finnish and Indian cultures are essential for the individuals, business leaders and multinational organizations involved in doing business from both the cultures. As a result, it is essential to cognize how Finland and India differ in their values and practices through the cultural typologies proposed by Hofstede and GLOBE study. In this scenario the aim of this study is to present the similarities and differences between Indian and Finnish cultures through the Hofstede and GLOBE dimensions. Therefore, the main research question of this study would be "What are the differences and similarities between Finnish and Indian culture that are identified in Hofstede and GLOBE study"?.

In this above discussed correlation this study is organized as follows. The first section of this study will present the Indian and Finnish culture through Hofstde's cultural dimensions. The second part of the study will discuss Indian and Finnish culture through GLOBE study. The third section of this paper will discuss the Indian and Finnish leadership style through GLOBE study. The fourth section of the study will discuss the major findings of the study and the final section of the study will present the conclusion of the study.

2. LITERATURE REVIEW

This study has taken Hofstede's cultural dimensions and GLOBE study for the following reason. The Hofstede's cultural dimension is the one of the first studies which opened up the discussion in comparing the organizational cultures of different nations. "Hofstede's observation and analysis provide scholars and practioners with a highly valuable insight to the dynamics of cross-cultural relationships". Moreover, "Hofstede's cultural dimensions has been instrumental in the implementation of many business systems including: compensation practice, budget control practices, entrepreneurial behavior, training design, conflict resolution, leadership styles, performance, and workgroup dynamics" (Micheael 1997, Smith 1998 in Jones, 2007:3). Likewise, the GLOBE study is the most recent survey which offers an analysis in differentiating the relation between the social values, social practices and leadership effectiveness of 62 societies or countries in the World (Schlosser, 2006; Terlutter, Diehl & Mueller, 2006). Moreover, the GLOBE study adds up new dimensions that were left in Hofstede and gives an overview about the preferred leadership of people from different parts of the World. Therefore, comparing the Indian and Finnish organizational cultures through the Hofstede and GLOBE study will offer novel insight in understanding the commonalities and distinction between Finnish and Indian organizational culture and preferred leadership styles.

2.1. Hofstede Cultural Dimensions

Hofstede's cultural dimensions help to look in-depth at the cultural differences between Finland and India. The Hofstede's cultural dimensions were predominantly constructed based on the data from the IBM employees based on 88,000 respondents in 20 languages from 66 countries (Hofstede, 2011; Terlutter. Diehl & Mueller, 2006). Hofstede proposed first four dimensions such as power distance, individualism vs. collectivism, masculinity vs. feminism and uncertainty avoidance. Later on Hofstede introduced the fifth dimension which is longterm orientation and the sixth dimension which is indulgence vs. restraint. The Hofstede's cultural dimensions are the most cited cultural typology which will be highly valuable in unveiling the dynamics of Indian and Finnish culture (Jones, 2007:2). Hofstede's proposed six cultural dimensions such as power distance index (PDI), Individualism vs. Collectivism (IDV), Masculinity vs. Feminity (MAS), Uncertainty avoidance index (UAI), long term vs. short term orientation (LTO) and indulgence vs. restraint (IVR) are explained and compared the scores of India and Finland in the following section. The following table gives the Hofstede's scores and ranks of Finland and India.

Table 1. Comparison of cultural dimension scores on India and Finland

CULTURAL DIMENSIONS	INDIA	RANK	FINLAND	RANK
Power distance index (PDI)	77	17-18	33	68
Individualism vs. Collectivism (IDV)	48	33	63	22
Masculinity vs. Feminity (MAS)	56	28-29	26	68
Uncertainty avoidance index (UAI)	40	66	59	50-51
Long term vs. Short term orientation (LTO)	51	40-41	38	51-54
Indulgence vs. Restraint (IVR)	26	73	57	27-29

Power Distance (PDI): Power distance is the degree to which the less powerful members of organizations and institutions (like the family) accept and expect that power is distributed unequally (Hofstede, 2010; 2011). The comparison of power distance (PDI) between both the countries reveals that India scores 77 in the power distance which is higher than the world average 56.5 and ranks between 17-18.Whereas, Finland stands on low power distance which scores 33 and it ranks 68 in the world scores (Hofstede,2010). It is evident from these scores to understand how hierarchical Indian society and work environment when compared with Finland. Power is distributed unequally in high power distance cultures like India as it is based on relationship and dependency. Likewise, in high PDI countries, less powerful members accept power relations that are more autocratic and paternalistic (Hofstede, 2010). Subordinates acknowledge the power of others simply based on where they are situated in certain formal, hierarchical positions (Hofstede, 2010). India is a culture with high power distance where the subordinates depend on the people with high status with power. The roots of hierarchy in India have been the replication of caste system that originated thousands of years ago and also due to the influence of migration and colonialism. The caste system in India has created a rigidness which created hierarchy between the members of its society. The caste system is a hierarchical ordering of society into groups that each has their own status, set of norms, and overall function such as priest, warrior, merchant, and worker (LeFebvre, 2011:4). The power distance between the members of the society has also transferred to workplaces which has created hierarchy in the organizations. For instance in Indian business organizations the decision making is made at senior level or by the top management. Moreover, in the organizations the junior members in the organizations are expected to give respect and display loyalty to their seniors in the organizations. Whereas, the low power distance culture like Finland the subordinates or the people have limited dependency with the people with high status and power. It is evident that Finnish organizations the power is distributed equally. Moreover, in Finnish organizations people are based on interdependence rather than relationship. The dependence on the superior is meager and the dependence relationship is mutual in low PDI cultures. Also, the relationship between the leader and subordinates is open which allows the subordinates to approaches their leaders easily (Paakkala, 2011:27). People relate to one another more as equals regardless of formal positions. Subordinates are more comfortable with and demand the right to contribute to and critique the decision making of those in power. Moreover, in low power distance culture where the decision making can be done even from a junior level without the consultation of the senior level people in the organizations. In the Finnish organizations the members are expected to behave equally where the gap between the members of the organization is minimal when compared with Indian organizations.

Individualism (IDV) vs. collectivism: Individualism refers to the societies or cultures in which the interest of individuals prevails over the interests of the group (Hofstede, 2010). "The individualism emphasis on individual goals, individual rights, autonomy, self-reliance, achievement orientation, and competitiveness" (Hofstede,2011). Collectivism on the other hand emphasis on collective goals, collective rights, interdependence, affiliation with the larger collective, cooperation, and harmony" (Kulkarni et al., 2010:95). The comparison between Individualism vs. Collectivism (IDV) dimensions Finland high on individualism which scores 63 and ranks 22. Whereas, India scores 48 and rank 33 which is considered as the collectivistic culture (Hofstede, 2010). In individualistic cultures like Finland people are expected to stand up for themselves and their immediate family, and to choose their own affiliations. The individualistic are focused on individual rights and personal achievement. Individualism pertains to societies in which the ties between individuals are loose. Management in an individualist society is management of individuals where the bonus and rewards are linked to individual performance. In Individual cultures the task is important over personal relationship (Hofstede, 2007:67). The relationship is playing a minimal role in establishing business relationship. In individualistic cultures like Finland business relationship are more on transactional basis when compared with collectivistic culture like India where the business relationship is viewed as long term and sustainable. Moreover, in collectivist societies such as India individuals act predominantly as members of a life-long and cohesive group or organization. In collectivistic societies the interest of the group members prevails over the interest of the individual which is seen in the families and also in the organizations. Moreover, in India people have large extended families, which are used as a protection in exchange for unquestioning loyalty. In the collectivistic cultures people are integrated in to cohesive group right from their birth and they protect their group throughout their life time. In collectivistic culture like India the relationship is prevailed over the task orientation. Moreover, in India personal relationship prevails over task relationship (Hofstede, 2003:47). Trust plays a major role in establishing business relationship as Indians tend to do business with the friends rather than with the unknown people. Besides, the individualistic cultures like Finland believes in individual accomplishments whereas the collectivistic culture like India is more concerned with group or team accomplishments. The social norm of individualistic culture is to have a nuclear and egalitarian family on the other hand the social norm in collectivistic culture is community family (Hofstede, 2005; Kazi, 2009).

Masculinity (MAS), vs. Femininity: Masculinity (MAS) vs. Femininity is the dimension that deals with the distribution of emotional roles between the genders within a culture (Hofstede, 2010). The gender roles are separated in masculine cultures which emphasis men to be assertive, tough and focused on material success and additionally, masculine cultures values competitiveness, assertiveness, materialism, ambition and power (Hofstede,2010). In the feminine cultures both women and men are supposed to be equal, modest, tender and concerned with quality of life (Hofstede,2010;2011). Feminine cultures place more value on relationship where men and women have the same values emphasizing modesty and caring (Hofstede, 2011:12). Hofstede's scores on the comparison between Masculinity vs. Feminity (MAS) dimensions reveals that India stands high on masculinity which scores 56 and ranks between 28-29. While, Finland scores 26 and ranks 68 which is considered to be feminine culture. The masculine culture like India value earnings, recognition, advancement and challenge. On the other hand Feminine culture Finland value personal relationships, care for others, quality of life and service (Francesco & Gold, 1998:18). The representation of women in workforce and representation of women in higher position in organization is lesser in Indian organizations as the culture is male dominated. It is obvious people display wealth and thirst for achievement in masculine culture and the feminine culture focus on preserving the environment and helped under privileged people. The perception such as "big is beautiful" is appreciated more in highly masculine culture whereas the feminine culture believes in the perception of "small is big" (Elkjaer et al, 2007: VII). The workplaces in masculine cultures prefer traits like authority, assertiveness, performance and success while the workplaces in feminine cultures tend to be more democratic (Jones, 2007:4).

Uncertainty avoidance index (UAI): This dimension deals with the society's tolerance for uncertainty and ambiguity. UAI can be defined as the extent of which the members of the society or culture feel threatened by ambiguous or unpredictable situations (Hofstede, 2010; 2011). The UAI cultures programs its members to feel either comfortable or uncomfortable in an ambiguous situation (Hofstede, 2011:10). In UAI dimension India scores 40 and ranks 66 and considered as weak uncertainty avoidance. But, Finland scores 59 and ranks 50-51 which is high in uncertainty avoidance. The cultures with high uncertainty avoidance try to minimize the occurrence of unusual circumstances by careful planning and implementing the rules and regulation to cope with ambiguous situations. The societies with low uncertainty avoidance are comfortable and ready to face the ambiguous situations and tend to be pragmatic and try to have few rules as possible which are based on realistic situations (Hofstede, 2010). High uncertainty cultures like Finland are highly structured with explicit rules of behavior and the law is respected seriously. The low uncertainty avoidance culture like India favors unstructured which makes it flexible for the people. Though there are strict laws exist it is not taken seriously as people are not completely aware of the legal aspects as it is not common to be aware of it (Hofstede, 2010; Francesca & Gold, 1998:18). People are threatened by the uncertain situations in high UAI cultures which lead them to create formal rules and believe in their correctness. The cultures that are weak in UAI the people are not threatened by the uncertain situations as uncertainty inherent in life is accepted and each day is taken as it comes. Therefore, weak UAI societies do not indulge in creating formal rules and believe in their correctness (Hofstede, 2011:10, Hofmann, 1999:19).

Long term orientation (LTO), vs. Short term orientation: First called "Confucian dynamism", it describes societies' time horizon (Hofstede,2010;2011). The LTO societies give more significance to the future which foster pragmatic values oriented towards rewards, including persistence, saving and capacity for adaptation. The short term orientation societies are concerned with the fostering of virtues related to the past and present-in particular, respect for tradition, preservation of face and fulfilling social obligations. The comparison between Long term vs. Short term orientation (LTO) dimensions reveals that– India scores 51 and ranks 40-41 which is high in long term orientation. Finland scores 38 and ranks 51 – 54 which is considered to be short term oriented society. According to Hofstede (2011:15) long term oriented culture like India will believe in the events that will occur in future and the short term oriented cultures like Finland believes in the important events that occurred in the past or takes place now. Deep sense of harmony and stable relationships are essential for doing business in LTO societies and personal steadiness and stability are required for short term oriented societies for the purpose of business (Hofmann, 1999:24).

Indulgence, vs. Restraint (IVR): "Indulgence stands for a society that allows relatively free gratification of basic and natural human desires related to enjoying life and having fun" (Hofstede, 2011). Restraint stands for a society that controls gratification of needs and regulates it by means of strict social norms" (Hofstede, 2011:15). The comparison between Indulgence vs. Restraint (IVR) dimension reveals that India scores 26 and ranks 73 which is considered as restraint society. Whereas, Finland scores 57 and ranks between 27-29 which is considered as highly indulgence society. The high indulgence society's people can freely fulfill their basic needs and desires freely as there are no strict social norms. On the other hand restraint societies have stricter social norms the gratifications of drives are suppressed and regulated (Hofstede, 2011:15). The indulgence society like Finland believes freedom of speech as important thing. Whereas, in the restraint society like India freedom speech is not a primary concern (Hofstede, 2011:15).

2.2. Globe Dimension

GLOBE stands for Global Leadership and Organizational Behavior Effectiveness. The GLOBE study used the results of previous empirical studies, other factors such as common language, geography, religion and historical accounts. Using the data collected on cultural values and beliefs from 62 countries and discriminant analysis (a technique which statistically test the extent to which GLOBEs classification is supported by the data) to confirm the clusters, globe developed a strong support to the existence of their proposed 10 cultural clusters (Elkjaer et al., 2007). Cultural similarity is greatest among societies that constitute a cluster, and cultural differences increases the farther clusters are apart. The GLOBEs ten clusters are Nordic Europe, Anglo, Germanic Europe, Latin Europe, Eastern Europe, Confucian Asia, South Asia, Arab and Sub-Saharan Africa (House et al, 2007). The GLOBEs nine cultural dimensions are power distance, uncertainty avoidance, humane orientation, institutional collectivism, in group collectivism, assertiveness, gender egalitarianism, future orientation and performance orientation. In this conceptualization GLOBE measures both cultural practice (the way things are) and values (the way things should be) at the organizational and social level of analysis. The GLOBE study is most recent study which offers an alternative perspective to the existing cultural dimensions such as Hofstede and Schwartz as it clearly distinguishes the social practice and values separately (Terlutter, Diel & Mueller, 2006:434). Besides, the GLOBE (2007) study also analyzed the preferred leadership styles of the 62 countries.

Uncertainty avoidance refers to the extent which a society, organization, or group relies on social norms, rules and procedures to alleviate unpredictability of future events (Javidan, Peter, Darfman, Luque & House et.al, 2006:70). This dimension emphasizes people's attitudes in seeking orderliness, consistency and structure (Javidan, Stahl, Brodbeck & Widerom, 2005:62). The societies that score high on uncertainty avoidance rely on formalized policies and procedures, establishing and following rules, verifying communications in writing and the societies that score low rely on informal interactions and informal norms rather than formalized policies, procedures and rules (House at.al. 2004). According to this dimension higher scores indicate greater uncertainty avoidance. In this dimension society practices (as is) indicate Finland scores 5.02 and India scores 4.15. The social values (should be) India scores 4.73 and Finland scores 3.85. The societal practices of this dimension reveal that Finnish society tends to avoid uncertainty when compared with Indian society. The Finnish culture seems to be less comfortable in handling uncertainty and the Indian culture is more relaxed in dealing with ambiguity.

Table 2. Comparison of society practice scores on India and Finland

GLOBE Cultural dimension	Society practice (As is) FINLAND	Rank FINLAND	Society practice (As is) INDIA	Rank INDIA
Uncertainty avoidance	5.02	8	4.15	29
Humane orientation	3.96	35	4.57	9
Institutional collectivism	4.63	10	4.38	25
In group collectivism	4.07	54	5.92	4
Assertiveness	3.81	47	3.73	53
Gender egalitarianism	3.35	31	2.90	55
Future orientation	4.24	14	4.19	15
Power distance	4.89	47	5.47	16
Performance orientation	3.81	46	4.25	23

Table 3. Comparison of social value scores on India and Finland

GLOBE Cultural dimension	Society value (Should be) FINLAND	Rank FINLAND	Society value (Should be) INDIA	Rank INDIA
Uncertainty avoidance	3.85	53	4.73	29
Humane orientation	5.81	2	5.28	44
Institutional collectivism	4.11	55	4.71	32
In group collectivism	5.42	47	5.32	50
Assertiveness	3.68	35	4.76	7
Gender egalitarianism	4.24	45	4.51	36
Future orientation	5.07	51	5.60	29
Power distance	2.19	60	2.64	38
Performance orientation	6.11	20	6.05	26

Humane orientation refers to the degree to which a collective encourages and rewards individuals for being fair, altruistic, generous, caring and kind to others (Javidan, Peter, Darfman, Luque & House et.al, 2006:70). The GLOBE study (2010) states that the societies that are with high humane orientation values altruism, benevolence, kindness, love and generosity. Whereas, the societies that have low humane orientation values of pleasure, comfort, and self-enjoyment (House at.al. 2004). According to this dimension higher scores indicate greater humane orientation. In this dimension in the society practices (the way things are) India scores 4.57 and Finland scores 3.96. The social values (the way things should be) scores indicate that Finland scores 5.81 and India scores 5.28. Historically Indian culture is deeply rooted with humane orientation as the most striking feature of ancient Indian civilization is its humanity (Chokkar, 2007:972). Moreover, Indian culture also believes in "Karma" which is the major belief in Hinduism means the action of an individual whether it is good or bad it come back to the individual in the future. Therefore, it is obvious India is highly humane oriented society which is also supported by the GLOBE scores which shows that the social practices of India scores higher on humane orientation than Finland.

Institutional collectivism refers to the degree which organizational and societal institutional practices encourages and reward collective distribution of resources and collective action (Javidan, Peter, Darfman, Luque & House et.al, 2006:70). This dimension emphasize the individuals encouragement by the society to be integrated into broader entities with harmony and cooperation as paramount principles at the expense of autonomy and individual freedom (Javidan, Stahl, Brodbeck & Wilderom,2005:62).The societies that are high on institutional collectivism tend to have members that assume that they are highly interdependent with the organization and believe it is important to make personal sacrifices to full their organizational obligations (House at.al.,2004). On the other hand the societies those scores high on individualism tend to have members that assume that they are independent of the organization and believe it is important to bring their unique skills and abilities to the organization (House at.al. 2004). According to this dimension higher scores indicate greater collectivism. In this dimension the societal practices (the way things are) indicate that Finland scores 4.63 and India scores 4.38. At the same time the scores of social values (the way things should be) India scores 4.71 and Finland scores 4.11.

In-Group collectivism refers to the degree which individuals express pride, loyalty and cohesiveness in their organizations or families (Javidan, Peter, Darfman, Luque & House et.al, 2006:70). The societies that are high on in group collectivism have individuals who are integrated in to strong cohesive groups and the societies that are low in this dimensions has the individuals who look after themselves or their immediate families (House et.al, 2004). According to these dimension higher scores indicates greater collectivism. In this dimension the societal practices (the way things are) India scores 5.92 and Finland scores 4.07. On the other hand the social values (the way things should be) Finland scores 5.42 and India scores 5.32. When comparing both the countries the scores of societal practices reveals that India scores very high on this dimension when compared with Finland though the social values predicts that Finnish society is more in-group collectivistic than India. India is a collectivistic society where the individuals are concerned about the common good of the group which they are associated with. Moreover, the collectivistic nature of Indians also embedded with the caste system as people are very much associated with their caste. It is also seen in the work place people work as teams where the people work for the common goal of the team. The roots of collectivism in India stem in part from the emphasis on family, caste system, sense of kinship, and community (Kulkarni et. al., 2010: 97). The family is the integral part of the collectivistic culture in India. For example in India quite often the families take decision on major issues of the individuals such as marriage (the family members choose the bride for their sons or daughters), education (family members advice what to study), job (family members advice on taking a job offer with a particular company) etc. It is also very common that people work in the teams hang out together even after their work. On the other hand in individualistic culture like Finland the role of family in individual's life is minimal and individuals are responsible for decision making in important events happen in their life.

Assertiveness refers the degree to which individuals are assertive, confrontational, and aggressive in their relationships with others (Javidan, Peter, Darfman, Luque & House et.al, 2006:70). According to GLOBE the societies that score higher on assertiveness value assertive, dominant, and tough behavior for everybody in society. These societies value competition which value success and progress (House at.al., 2004). On the other hand the societies that score lower on assertiveness tend to view assertiveness as socially unacceptable and value modesty and tenderness. Moreover, this societies value cooperation which has value for people and relationships. According to GLOBE study the society practice (the way things are) reflects that Finland scores 3.81 and India scores 3.73. In the social values (the way things should be) India scores 4.76 and Finland scores 3.68. In this dimension Finland scores high according to the social values (as is) when compared with India. The social values of GLOBE shows that India is high on assertiveness which values competition, success and progress and on the other hand Finland scores low on assertiveness which values cooperation, warm and relationships (Grove,2004 & House,2004).

Gender egalitarianism refers to the degree to which collective minimizes gender equality (Javidan, Peter, Darfman, Luque & House, 2006:70). According to this dimension the lower scores indicates greater male domination and the lower dimension indicate gender equality (House et.al, 2004). In this dimension based on the society practices (the way things are) Finland scores 3.35 and India scores 2.90. Finland has more gender egalitarianism than India. In the society values (the way things should be) India scores 4.51 and Finland scores 4.24. In the societal practice India has greater male domination. According to GLOBE the societies that score high on gender egalitarianism have more women in the positions of authority and as well as high status in the society (House et.al, 2004). Whereas, the societies that score lower on gender egalitarianism tend to have fewer position of authority and will have women in lower status in the society (House et.al, 2004). When comparing the societal practices between both the countries India scores low as it is more male dominated society where men tend to be dominant. Finland scores higher on this dimension where the women have equal value so as man in the organizations. Though scores on the social values reflects that India scores high on this dimension when compared with Finland it is assumable the role of women in the society have gained importance over a period of time.

Future orientation refers to the degree to which individuals engage in future-oriented behaviors such as delaying gratification, planning, and investing in the future (Javidan, Peter, Darfman, Luque & House et.al, 2006:70). The societies that are high on future orientation tend to achieve economic success whereas, the societies lower on future orientation tend to have lower economic success (House et.al, 2004). The scores of this dimension based on society practices (the way things are) Finland scores 4.24 and India scores 4.19. The scores of society values (the way things should be) shows that India scores 5.60 and Finland scores 5.07. The future orientation in Finland denote systematic planning process, careful in risk taking and opportunistic decision making (Javidan, Peter, Darfman, Luque & House, 2006:69). On the other hand future orientation is perceived differently in India which tends to be more based astrological predictions. In India people go for astrologers to ask for predictions on several activities such as employment, marriage, wellbeing, health issues, starting new ventures and other activities. The Indian astrologers offer astrological predictions for the people to give remedies solve problems and predict the future of people's life based on the planet positions and astrological calculations. Besides, people also look for "Vastu" predictions to build their houses and official buildings which is similar to Chinese "Feng Shui". Moreover, people also look for auspicious time to start new activities as the activities they start should be successful in long-term. The GLOBE scores reveals that the societal practices states that Finland scores high on future orientation and the societal values remains contradictory.

Power distance is the degree to which members of an organization or society except and agree that power and privilege should be stratified and concentrated at higher levels of an organization (Javidan, Peter, Darfman, Luque & House, 2006:70). According to GLOBE study in the high power distance culture power is seen as providing social order, relational harmony and role stability whereas the low power distance cultures considers power as source of corruption, coercion, and dominance (House eta al...2004:536). In this dimension based on society practices (as is) India scores 5.47 and Finland scores 4.89.The scores of social values (should be) India ranks 2.64 and Finland scores 2.19. From this scores it is evident India is high power distance culture when compared with Finland. Finnish organizations are less hierarchical where the individual employees are encouraged to take independent decisions. Moreover, the Finnish managers do not under estimate their employee's capabilities to make decisions (Elkjaer, 2007). On the other hand Indian culture is hierarchical where the decision making is most often taken by the managers or the senior most people. Moreover, the employees or subordinates in a team are obliged to agree with their managers and they may not have an opportunity to disagree. Besides, the subordinates expect the superior to be benevolent towards them, and if this benevolence is reciprocated, the subordinate is also likely to respond by remaining loyal (Kumar, 2005:3). From this it is understandable that Finland is low power distance culture and India is high power distance culture.

Performance orientation refers to which the collective encourages and rewards group members for performance improvement and excellence (Javidan, Peter, Darfman, Luque & House et.al, 2006:70). Moreover, it also reflects to the extent which a community encourages and rewards innovation, high standards, and performance improvement (Javidan, 2004:239). However, according to GLOBE the societies that scores higher on performance orientation emphasize results more than people. Moreover these societies value assertiveness, competitiveness and materialism. Whereas, the societies that scores lower on performance orientation emphasize loyalty and belongingness. Besides, these societies also value harmony with environment rather than control. In this dimension the scores of society practices (the way things are) of India is 4.25 and Finland scores 3.81. In this aspect India ranks higher in performance orientation than Finland. In the aspect of society values (the way things should be) Finland scores 6.11 and India scores 6.05. The social value scores of India and Finland are closer to each other which mean both the countries societal values are based on performance orientation. The data shows that the Finnish society has more emphasize towards performance than the Indian society. As there is contradiction between the scores of both the countries in society practice and social values it is hard to generalize the fact about the performance orientation. The below mentioned graphs explain the comparison of Indian and Finnish social value and social practice cultural dimensions.

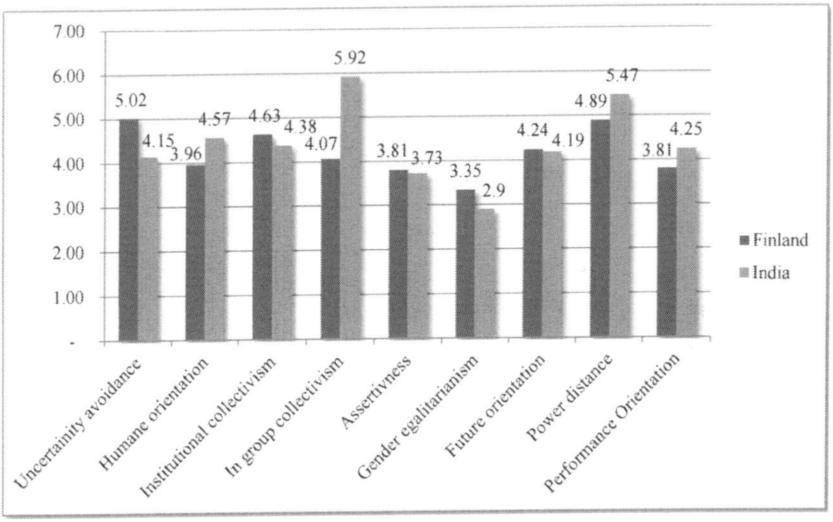

Figure 1. Comparison of society practice scores on India and Finland

The social practice scores of Finland shows that the top scores of dimensions are high on uncertainty avoidance (5.02), power distance (4.89) and institutional collectivism (4.63). The social practice scores of India reveals that the top scoring dimensions are in-group collectivism (5.92), power distance (5.47), and humane orientation (4.57).

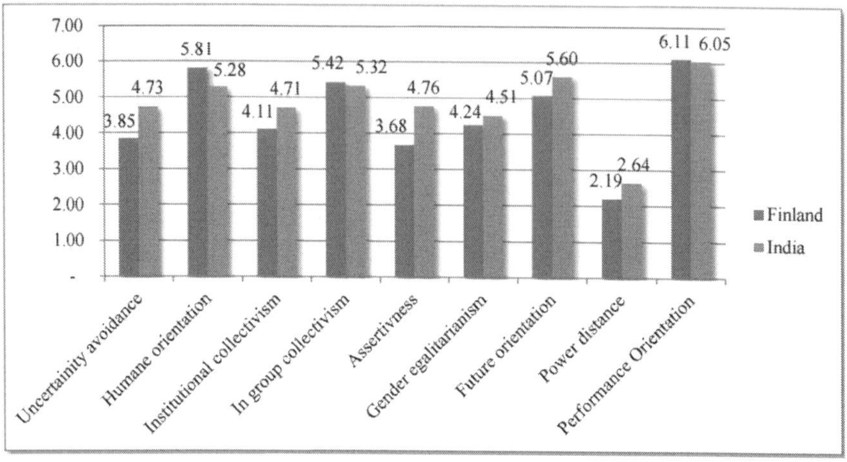

Figure 2. Comparison of society values scores on India and Finland .

The social value dimensions of Finland reveals that the top ranking dimensions are performance orientation (6.11), humane orientation (5.81) and in group collectivism 85.42).

The social value dimensions of India reveals that the top ranking dimensions are performance orientation (6.05), future orientation (5.60) and in group collectivism (5.32).

2.3. GLOBE Leadership styles

The GLOBE study also presented its research on leadership dimensions viewed by the people of the countries that have been researched. According to GLOBE people from different cultural groups share a high level of agreement on their beliefs about effective leadership and that significant statistical differences exist among cultural groups in their beliefs about leadership (House et.al, 2004). This shared belief may be described as culturally endorsed implicit theory of leadership (CLT). GLOBE study analyzed the responses of 17,300 middle managers and identified the 112 leadership characteristics. The analysis of the study generated 21 leadership scales which were statistically and conceptually reduced to six scales which resulted leadership styles such as performance-oriented style, team-oriented style, participative style, humane style, autonomous style and self-protective style. The following table shows the preferred leadership style in Finnish and Indian cultures according to the GLOBE study.

Table 4. Comparison of preferred leadership styles in India and Finland

LEADERSHIP STYLES	FINLAND	INDIA
Charismatic / value based	5.94	5.85
Team oriented style	5.85	5.72
Participative style	5.91	4.99
Humane oriented style	4.30	5.26
Autonomous style	4.08	3.85
Self-protective style	2.55	3.77

Charismatic / value based leadership style reflects the ability to inspire and motivate and achieve high performance from their followers on the basis of firmly held core beliefs (Javidan, Darfman, Luque & House, 2006:73). Charismatic leadership style creates a passion among the followers to perform better as they are motivated by the core values of the leader whom they follow. This dimension includes six leadership sub scales such as visionary, inspirational, self-sacrifice, integrity, decisive, and performance oriented (House at.al, 2004). The comparison between the preferred leadership style of Indians and Finnish managers shows that the preference of charismatic leadership or value based leadership is higher in Finland (5.94) than in India (5.84). Charismatic leadership style is the most preferred style both in Finland and India. Indians scores lower in this style when comparing Finland however, the notion of being charismatic have different meaning between Indians and Finns. India has seen many great leaders like Mahatma Gandhi, Subash Chandra Bose, Swami Vivekananda, and Jamsedji Tata whom have been considered as role model and visionaries. Indian subordinates view their leaders as the one who inspire and motivate them. Indian subordinates look at the leaders as a role model (the one who behaves ideally as a good leader in personal life as well as the public life) and the one who cares about them by bringing integrity to work for the common goal of the company. It could be taken that Indian leaders are viewed by the subordinates as the one who is transformational who creates a common bonding with them. On the other hand the Finnish leaders are viewed by their subordinates as the one who provide advice and support, provide feedback on successful work and show confidence with them (Elkjaer et al. 2009).

Team-oriented style refers effective team building to achieve the goal in the organization (Javidan, Darfman, Luque & House, 2006:73). Team building style instills pride, loyalty, and collaboration among organizational members; and highly values team cohesiveness and a common purpose or goals (House et.al, 2004). This globe CLT dimension includes five primary leadership scales such as collaborative team orientation, team integrator, diplomatic, malevolent and administratively competent. The comparison between Finland and India show that the preference for team oriented style is higher in individualistic culture Finland (5.85) when compared with the collectivistic India (5.72). In the Finnish work environment the Finnish leaders should set clear goals and define objectives and each of the team members have an important role to play and each of the members are expected to contribute in their unique way (Lewis, 2007 in Elkjaer et al. 2009). Moreover, in the Finnish team oriented style the team members have the liberty to be independent and come up with their own decisions. On the other hand team oriented style in India works in different way where the team leaders specify the task for each team members and the team members are expected to perform by the given orders. Often the team member's motivation is to be loyal to the team leader and follow the orders given by the team leaders. Independent decision making may not be common among the team members. The main aim of the Indian leader is to get the work done efficiently through his team members who are clearly monitored. Moreover, it is also not common the team members outperform their bosses in the Indian environment. Challenging and outperforming the leaders or bosses may have serious consequences for the team members in the near future (Pal & Kapur, 2011). Though Finland and India both are closer in preferring team oriented leadership style it works differently in both the countries. India is a high power distance culture which is hierarchical in nature. Whereas, Finland is low power distance culture which has flat hierarchical structure. From this it is obvious the team oriented style may have different style in both the countries.

Participative style reflects the style which includes the managers and the employees involve in formulating and implementing decisions (Javidan, Darfman, Luque & House, 2006:73). Participative style encourages input from others in decision-making and implementation: and emphasizes delegation and equality which includes two primary leadership subscales such as autocratic and non-participative (House et.al, 2004). The comparison between the scores of Finland and India shows that Finnish employees and managers preference for participative style is higher (5.91) when compared with the Indians (4.99). India scores low in participative leadership style as it is a high power distance culture it also creates a business environment where the authority lies with the boss and the employees are generally not empowered. The employees are expected to follow their leaders even the employees may know that their boss is incorrect (LeFebvre, 2011; Kumar, 2005). Therefore, most often the decision making is done by the senior member of the Indian organization in which the other members of the organization have minimal role in decision making. Moreover, the employees do not expect them to play a role in decision making which show that authoritarian style of decision making is common. Therefore, though the subordinates are well educated and knowledgeable than their leaders still they are expected to behave in a submissive way by not challenging or surpassing their bosses. In this kind of traditionally rooted leadership style though the Indian leaders are independent and authoritative and therefore participative leadership is not common among the Indian organizations. The Finnish leadership style is democratic approach which allows the employees to work with their leaders and make decisions flexibly together with their leaders (Vesterinen, Isola & Paasivaara, 2009:506). Therefore, participative leadership style is more common in Finnish work organizations when compared with Indian one.

Humane oriented style is about being supportive and considerate leaders to the managers and employees with generosity and compassion (Javidan, Darfman, Luque & House, 2006:73). Humane style stresses being supportive, compassionate, modes, generous and concerned with the well-being of others which includes two leadership subscales such as modesty and humane oriented (House et.al, 2004). Indians employees are more positive towards humane oriented style (5.26) when compared with Finns (4.3) who ranks lower. India scores high on this as the dimension as the Indian leaders give importance to various societal activities. Indian business environment is uncertain, dynamic and extremely fluid. Moreover, the Indian business environment also has poor infrastructure with various challenges (Useem, 2010:23). It is extremely important for the Indian leaders to cultivate a good image and reputation through their wealth. Therefore, the Indian leaders involve in nation building activities and also various philanthropic activities. By doing these activities Indian leaders are also seen as motivators and role models. Whereas, Finland is a highly developed society with good infrastructure which may not require their business leaders to involve in social and philanthropic activities and also Finnish traditional values are based on modesty where displaying ones wealth may be considered as arrogant (Elkjaer, 2009).

Autonomous style is characterized by an independent, individualistic, and self-centric approach to leadership which includes one primary leadership scale which is labeled as autonomous (Javidan, Darfman, Luque & House, 2006:73). The comparison between the scores of Finland and India shows that Finnish employees and managers prefer more autonomous style (4.08) when compared with Indians (3.85). Lindell & Sigfrids, (2007) in Elkjaer et al. (2009) states that the Finns are highly educated and knowledgeable which means the subordinates who work under the Finnish leader are also the same. Moreover, according to the authors it is up to the leaders to inspire and get the best out of their employees. So in case of Finland also the subordinates who work under the leader have the same knowledge or more knowledge than the leader which is hard to dictate them in an authoritative manner. The Indian leadership style is rooted in the traditional idea of benevolent paternalism which means that the bosses have the right to order at the same time take care of his subordinates (Arora, 2010:21).Moreover, India is a collectivistic culture therefore, it requires the Indian leaders to behave in a socially accepted way. Being autonomous, independent, individualistic and selfcentered may be considered as selfish or not caring in Indian environment. Whereas, Finland is an individualistic culture being independent and self-centered is socially accepted behavior therefore, autonomous leadership style is more common in Finland than in India.

Self-protective style emphasizes procedural, status-conscious, and face-saving behaviors and focuses on the safety and security of the individual and group (Javidan, Darfman, Luque & House, 2006:73). This leadership dimension includes five primary leadership subscales such as self-centered, status conscious, conflict inducer, face saver and procedural (House et al., 2004). The comparison between Finland and India shows that Self-protective style is preferred higher by the Indians (3.77) when compared with Finland (2.55). Due to collectivistic culture in India the power distance is higher between the Indian leaders and their followers. The subordinates are expected to be give respect to their superiors, be obedient, submissive, and be formal with their leaders. Though there is this kind of expectations there is much emphasis is placed on the quality of relationship between the leader and his subordinates. Therefore, the Indian leaders are status conscious and exhibit face saving behaviors as the Indian leaders are expected to behave as a father in a family who takes care of everybody's needs. Paternalism is considered as one of the traits for a good leader.

Therefore, self-protective leadership is essential for the Indian business leaders to maintain the harmony and to be procedural in their behavior. On the other Finland is an egalitarian society and the status conscious are low among the members of the society and the concept of face saving does not have much emphasize in the Finnish society. Moreover, as Finland is a highly organized state the security of the individual and group in the organizations does not required to be taken care by the state. So there less expectation on self-protective style among the Finnish employees.

3. RESULTS AND DISCUSSION

This study is intended to present the differences and similarities between Finnish and Indian culture through the cultural dimensions proposed by Hofstede and GLOBE study. The findings of the study are as follows.

Hofstede

The scores or rank on Hofstede's (2010) dimension on India shows that out of the six dimensions India scores high on power distance which is followed by other dimensions such as long term orientation, masculinity, collectivism, uncertainty avoidance and indulgence. On the other hand Finland scores high on Individualism, followed by uncertainty avoidance, indulgence, long term orientation, power distance and masculinity (Hofstede, 2010). From the comparison it is obvious to understand the wide gap in the cultural differences between Finland and India. Hofstede's, (2010) six cultural dimensions may not be the perfect indicator in giving the exact information about both the cultures. However, these dimensions will provide a basic understanding on the national cultures between both the countries. Moreover, analyzing the cultural dimensions is an endeavor to understand the Finnish and Indian culture as a large group at the general level. The results of these cultural dimensions may be contrasting at Individual level as the individual personalities from both the culture may have different values and behavior. The below mentioned graph explain the comparison of Indian and Finnish cultural dimensions.

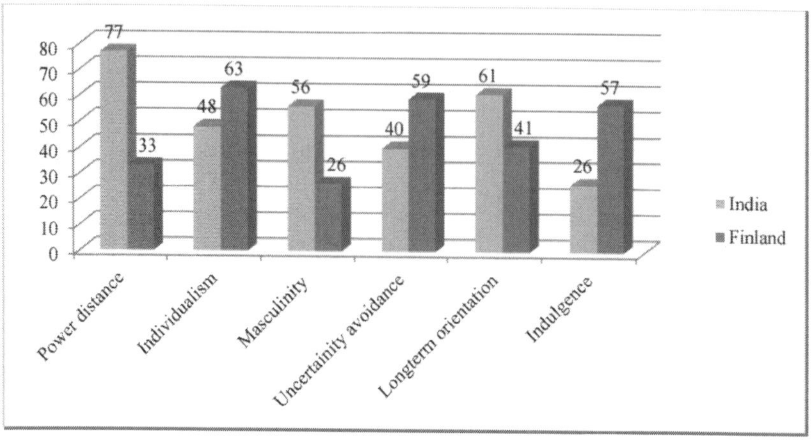

Figure 3. Comparison of cultural dimension scores on India and Finland

GLOBE

The top three ranks of Indian social practice GLOBE cultural dimensions are in-group collectivism (4), humane orientation (9) and future orientation (14). The top three ranks of Finnish social practice cultural dimensions are uncertainty avoidance (8), institutional collectivism (10) and future orientation (14). The top three ranks of Indian social values are assertiveness (7), performance orientation (26), uncertainty avoidance (29) and future orientation (29). The top ranking Finnish social values are humane orientation (2), performance orientation (20) and assertiveness (35) (House et al., 2004). The below mentioned graph explain the comparison of Indian and Finnish cultural dimensions.

Hofstede and GLOBE

The findings of the study of Hofstede and GLOBE study reveal that there is big difference in hierarchy between the Finnish and Indian culture. The Hofstede score on power distance (PDI) for India is 77 and Finland is 33 and the different between both the cultures is 44 which is a major difference. Therefore, it is assumable that power distance is greater in Indian culture when compared with the Finnish culture. The low power distance of Finland is also visible both in Hofstede and GLOBE study. As per the GLOBE social practice India scores 5.47 and Finland scores 4.89 and the difference is 0.58. The high power distance in India is visible in both Hofstede and GLOBE study. The Hofstede score on individualism vs. collectivism (IDV) dimension is that Finland scores 63 on individualism and India scores 48 in this dimension and the difference between both the cultures is 15. Hofstede scores confirm Finland is more individualistic than India. As per the GLOBE India scores 5.92 and Finland scores 4.07 and the difference between both the cultures is that 1.85. It is evident from this scores that the GLOBE study also confirms the Indian social practices are high on in group collectivism when compared with the Finnish culture. Therefore there is a huge difference in handling power relation in both the cultures.

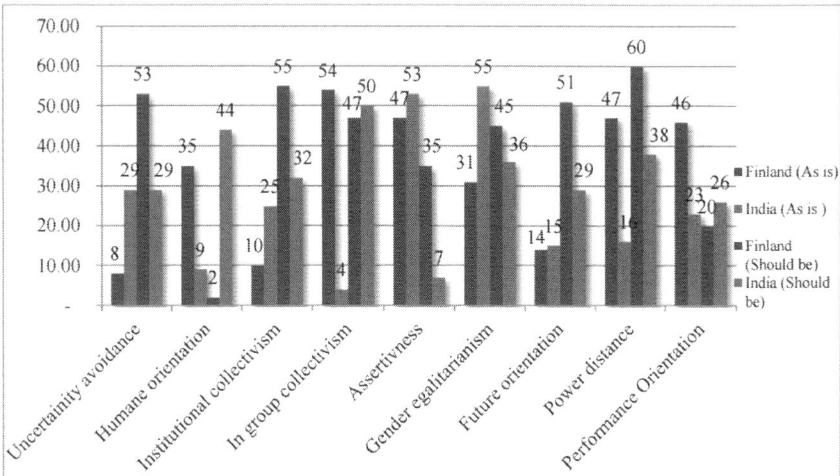

Figure 4. Comparison of society practice and society values scores on India and Finland.

In the masculinity vs. feminity (MAS) dimension India scores 56 and Finland scores 26 and the difference between both the cultures is 30. This confirms that India is high masculine culture when compared with India. In correlation with MAS dimension the GLOBE score on society practice of gender egalitarian scores of Finland is 3.35 and India is 2.9 and the difference is 0.45. The Hofstede score on uncertainty avoidance index (UAI) for Finland is 59 and India is 40 and the difference between both the cultures is 19. The GLOBE social practice scores of uncertainty of Finland is 5.02 and India is 4.15 and the difference between both the cultures is 0.87.

The Hofstede (2010) score on long term vs short term orientation (LTO) for India is 51 and Finland is 38 and the difference between both the cultures is 13. The Hofstede (2010) score confirms Indian culture is more long term oriented than the Finnish one. However, the GLOBE score on society practice on future orientation of Finland is 4.24 and India is 4.19 and the difference is 0.05. The results of GLOBE (2004) shows Finland is more future oriented society than India by little difference which can interpret as both Finland and India are future oriented society.

The Hofstede (2010) score on indulgence vs. restraint (IVR) for Finland is 57 and India is 26 and the difference between both the cultures is 31 which confirm Finland is an indulgence society and India is a restraint society. The GLOBE score on society practice of institutional collectivism for Finland is 4.63 and India is 4.38 and the difference is 0.25. The GLOBE (2010) score on society practice of performance orientation is higher in India with 4.25 and Finland is lower with 3.81. The difference between both the cultures is 0.44 which confirm that performance is high in India than in Finland. The GLOBE score on society practice of assertiveness is higher in Finland with 3.81 and India is 3.73 and the difference between both the cultures is that 0.08. The GLOBE scores confirm that both Finland and India is assertive societies.

GLOBE Leadership styles

In the GLOBE study India belongs to south Asian cluster which prefers charismatic or value based leadership style (5.85) the most. The second preferred leadership style is team oriented style (5.72) and the humane oriented style (5.26) is viewed higher and comes as the third preferred leadership style. Participative leadership style (4.99) comes in the fourth position which is followed by autonomous style (3.85) and self-protective style (3.77) as fifth and sixth preferred leadership style among the Indians (House et.al, 2004). The scores of all preferred leadership style in India are lesser when compared with the scores of South Asian cluster. Whereas, Finland comes under the Nordic cluster in the GLOBE which prefers charismatic or value based leadership as the most preferred leadership style. Charismatic leadership style is higher in Finland (5.94) when compared with the average (5.93) of Nordic cluster. Participative style comes as the next or second most preferred leadership style in Finland. The scores of participative leadership style are higher in Finland (5.91) when compared with the average (5.75) of Nordic cluster. Team oriented style is the third most preferred leadership style in Finland. Humane oriented style, autonomous style and selfprotective style are preferred fourth, fifth and sixth respectively

(House et. al, 2004). The below mentioned graph explain the leadership preference among Finnish and Indian employees.

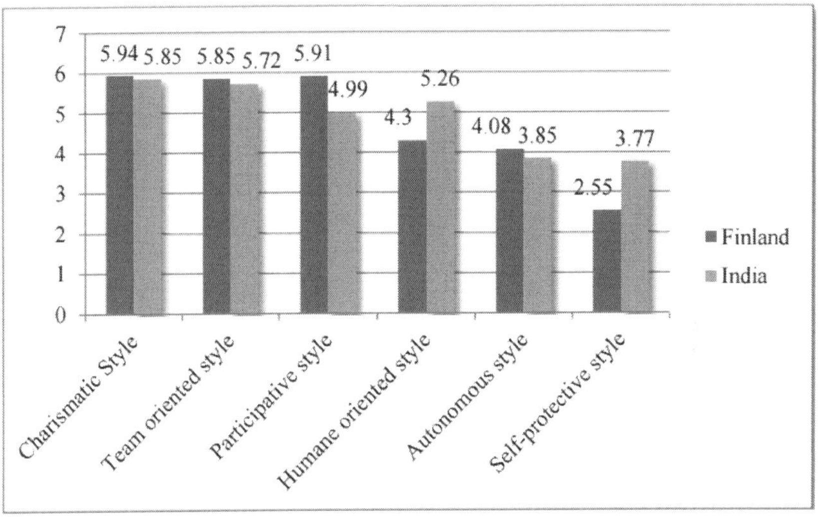

Figure 5. Comparison of preferred leadership styles in India and Finland

The preference for the leadership among the Finnish and Indian employees has both similarities and differences. The similarity is that both India and Finland scores high on charismatic leadership style. Though both Finland and India are very different culturally the preference for charismatic leaders is high in both the cultures. This shows that irrespective of cultures the employees prefer leaders who can inspire and motivate the employees to achieve performance from them. Finland scores 5.94 for the preference of charismatic style and India scores 5.85. The difference between both the cultures is that 0.09 which shows that there is more similarity in the preference between both the cultures even though there is no complete similarity between both the cultures. Another similarity between both the India and Finland is that the preference for team oriented style is high among both Finnish and Indian employees. Finland scores 5.85 and India scores 5.72 and the difference between the preference among the employees of both the cultures is 0.13. Therefore, the preference is very closer among both the cultures.

However, the preference of other leadership style among both the cultures is vast. For example, the preference for participative leadership style according to GLOBE study is higher in Finland than compared with India. Finland scores 5.91 and India scores 4.99 and the difference between both the cultures is 0.92. The participative leadership is preferred more among the Finnish employees as there may be the reason due to the power distance. Finland is a small power distance culture the use of power should be legitimate and is subject to criteria of good and evil (Hofstede,2011:9). Therefore, the inequality between the leaders and employees are less and the employees are encouraged and motivated to take part in the decision making process. Whereas, in India the power distance is large and the legitimacy of power is irrelevant as power is the basic fact of the society (Hofsetde, 2011:9). Therefore, there is inequality exist between the power relations of employees and the leaders and the employees are not encouraged to participate or consulted in decision making. The difference in preference for humane oriented style is higher among the Indian employees than the Finnish employees. India scores 5.26 and Finland scores 4.3 and the difference is 0.96. This also confirms the huge difference between both the cultures. The Indian culture believes in the notion of "Karma" which means the individuals are responsible for their own actions. It means if people do good it comes back and if they do badly it comes as well which simply denotes what goes around comes around. Due to the influence of this belief may also be one of the reasons that India scores on Humane oriented style. On the other hand such belief may not be existing in Finnish culture which may also be one of the reasons for low preference in humane oriented leadership style among the Finnish employees.

Besides, the difference in preference for autonomous style is slightly higher among the Finnish employees than the Indian employees. Finland scores 4.08 and India scores 3.85 and the difference is 0.23. The reason for the preference for autonomous style among the Finnish employees may also be due to the individualistic culture. In the individualistic culture like Finland the individuals are expected to take care themselves and the individuals are "I" conscious (Hofstede, 2011:11). Therefore, the people look at others as individuals who also reflect their expectation and preference regarding the leaders who are independent and selfcentered. Whereas, in the collectivistic culture like India the Individuals are expected to look after their siblings and the individuals are "we" conscious (Hofstede, 2011:11). Therefore, people expect their leaders to be paternalistic which also reflects their low preference for autonomous style. The comparison of preferred leadership style also reveals that there is a huge difference between the preference of self-protective style among the Finnish and Indian employees. India scores (3.77) and the Finland scores (2.55) which reveals that the high power distance and collectivistic culture is replicated in the self-protective leadership style. As Finland is small power distance and individualistic culture the need for face saving behavior and status consciousness are lower. Therefore, self-protective leadership style is not the most preferred one among the Finnish employees.

4. CONCLUSION

The in depth review of Finnish and Indian culture through the Hofstede and GLOBE study shows that there are significant differences and similarities exist between both the cultures. The review reveals that there is massive difference in terms of power distance, gender egalitarianism, individualism uncertainty avoidance, future orientation, uncertainty avoidance etc. At the same time there are also similarities in terms of leadership style such as charismatic and team oriented style. However, it is still questionable whether the similarities on the cultural dimension have same meaning in both the cultures. Moreover, generalizing Finnish culture through the cultural dimensions proposed by Hofstede and GLOBE study is more rational as Finland is more homogenous in nature with approximately 5.4 million people as its population. However, generalizing India may not be more rational due to its huge population (approximately 1.27 billion people), cultural pluralism, and diversity. Besides, Indian cultural values, beliefs and conditions are contradictory from state to state and region to region (Sebastian, Parameswaran & Yahya 2006:1). Therefore generalizing Indian culture can lead to confrontation due to its history and geography etc. Rather than thinking India as one single culture it is wiser to think it as an association of various regional and subcultures. The contribution of this study to the literatures is as follows. First, this study will be one of the first study which gives insight by comparing the similarities and differences between Finnish and Indian organizational culture and preferred leadership style. Second, this study will be one of the few studies that compare Indian and Finnish organizational culture through Hofstede and GLOBE study. Third, this study is also one of the few studies that compare the countries which are totally dissimilar in terms of population. Besides, the study also offers several important managerial implications. First, the study explains the commonalities and distinction between Finland and India. Second, this study explains the power relations, gender egalitarianism, future orientation, uncertainty avoidance, institutional collectivism between both the countries. Third, this study explains the preferred leadership style among the employees of Finnish and Indian organizations. Fourth, this study gives an overview to the Finnish and Indian organizational cultures. The findings of the study will be highly practical for the Indian and Finnish business leaders who are involved in doing business in Finland and India.

This study has significant limitations. This study is a review study investigating the Finnish and Indian culture through empirical research could shed light on more interesting facts on Finnish and Indian culture. In order to understand Indian culture completely it is significant to conduct empirical study with focus on different regions, zones and states in India. At the same time conducting a similar empirical study on different regions of Finland may also reveal more insights to the Finnish culture. To conclude the intention of this study in general to give a bird's eye view of Finnish and Indian organizational culture and leadership style through Hofstede and GLOBE study.

REFERENCES

1. Arora, D., 2005., Foreign Multinational in India: Adapting to India's work culture and management practices. Internationale Geschäftstätigkeit in Asien, Munich, Rainer Hampp Verlag.

2. Ang, S., Dyne, L.V., and Tan,M.L., 2011., The Cambridge handbook of intelligence- Cultural Intelligence. Cambridge university press, UK. ISBN 978-0-521-51806-2

3. Chhokar, J.S., Bordbeck, F.C., House,R.J., 2008., Culture and leadership across the World. The GLOBE book of in-depth studies of 25 societies. Psychology press. Taylor and Francis group, USA. ISBN 13: 978-0-8058-5997-3

4. Chhokar,J.S., (2000) Leadership and culture in India: The GLOBE research project. Davis,K., and Newstrom,J.W., 1989., Human behavior at work. –Organizational behavior. McGraw Hill, USA. ISBN0-07-015574-7.

5. Drogendijk,R., and Zander,L., 2010., Walking the cultural distance: In Search of direction beyond friction. The past, present and future of International business and management advances in International management. Volume 23,189-212, Emerald publishing group, ISSN 1571-5027

6. Edinger,P., 2011., Relationships between culture and personality structures. Master Thesis. University of Vaasa, Finland.

7. Elkjaer,J.K., Trabolt,K.L., Hoie,T., Lindell,M., and Swenson,D., 2009., A Global Leadership Country report-Introduction. Copenhagen Business School, Denmark.

8. Francesco, A.M., and Gold,B.A.,1998., International Organizational behavior-Text, readings, cases and skills. Prentice hall.USA, ISBN0-13-192485-0.

9. Fons Trompenaars and Charles Hampden-Turner (1998) Riding the waves of culture. McGraw Hill. USA. ISBN 0-7863-1125-8.

10. Fougere, M.,2004., Finnish-French fundamental cultural antagonisms in organizing. Swedish School of Economics and Business Administration. Finland. ISBN 951-555-833-6.

11. Gamst,F.C., and Norbeck,E., 1976., Ideas of Culture. USA. Holt McDougal, ISBN 10:0030158664.

12. Grove,C.N., 2013., Worldwide differences in business values and practices: Overview of GLOBE research findings. Grovewell leadership solutions LLC.

13. Goel,P., 2011., Organizational differences in India. POMS 22nd Annual Conference. Reno, Nevada,USA. April 29th to May 2 nd , 2011.

14. Gupta,V., Surie,G., Javidan,M., and Chhokar,J., 2002., Southern Asia cluster: where the old meets the new?. Journal of World Business, 37 (2002) 16-27.

15. Gupta,V., Hanges,P.J., and Dorfman,P., 2002., Cultural clusters: methodology and findings. Journal of World Business, 37 (2002) 11-15

16. Hofmann,O., 1999., Differences in International Business culture.

17. Hofstede, G., 1997., Cultures and Organizations: Software of the mind, New York, McGraw Hill, ISBN 0077074742.

18. Hofstede,G., 1991., Cultures and Organizations: Software of the mind, Macmillan, London. Hofstede,G., 2010 Cultures and Organizations: Software of the mind. McGraw Hill, 3rd edition. ISBN-100071664181, ISBN-13:978-0071664189

19. Hofstede,G., 2011., Dimensionalizing Cultures: The Hofstede Model in Context. International Association for cross-cultural psychology, ISBN 978-0-9845627-0-1

20. House,R.J., Hanges.P.J., Javidan.M.,,.Dorfman.P.W., and Gupta.V., (2004) Culture, Leadership and Organizations-The GLOBE study of 62 societies. Sage Publications, USA. ISBN978-0-7619-2401-2

21. Javidan,M., Dorfman,P.W., Luque,M.S.De., and House ,R.J., 2006., In the eye of the beholder: Cross Cultural Lessons in Leadership from project GLOBE. Academy of Management perspectives.

22. Javidan,M., Stahl,G.K., Brodbeck,F., and Wilderom,C.P.M., 2005., Cross-border transfer of knowledge: Cultural lessons from Project GLOBE. Academy of Management Executive, Vol 19, No 2.

23. Jones,M.L.,2007., Hofstede-Culturally Questionable? Oxford Business & Economics Conference. OxfordUK,24-26 June 2007.

24. Kazi,S., 2009., Managerial Decision making style: India, Bangladesh, and Finland. Cultural side of the game. Tampere University of Applied Sciences, Finland.

25. Kulkarni,S.P.,Hudson,T.,Ramamoorthy,N.,Marchev,A.,Kondakova,P.G. ,and Gorskov,V., 2010., Dimensions of individualism–collectivism: A

comparative study of five cultures. Current Issues of Business and Law, Vol.5, ISSN 1822-9530

26. Kumar,R.,2005., Negotiating with the complex, Imaginative Indian. Ivey Business Journal, Ivey Management services, London

27. Lamsa,T., 2010., Leadership styles and decision making in Finnish and Swedish organizations. Review of International Comparative Management, Volume 11, Issue 1, March 2010.

28. LeFebvre .R. 2011., Cross-cultural comparison of business ethics in the U.S. and India: A study of business codes of conduct. Journal of emerging knowledge on emerging markets, Volume 3, Nov2011. India, China and America Institute. The Berkley electronic press.

29. Lewis,R.D., 1999., Cross-cultural communication - A visual approach. Great Britain. Transcreen publishing, ISBN 0953439802.

30. Ng, K.Y., Dyne,L.V., and Ang,S., 2009., Beyond International Experience: The strategic role of cultural intelligence for executive selection in IHRM.-Handbook of International Human resource Management, John Wiley & Sons limited.

31. Oinas,P., 2005., Finland: A success Story? Taylor and Francis, ISSN 0965-4313 print/ ISSN 1469-5944

32. Overgaard,L., 2010., An analysis of Indian culture in an era of globalization. Department of languages and business communication. Aarhus school of business. Aarhus University, Denmark.

33. Paakkala.P., 2011., Exploring corporate culture and communication in Finnish and Italian business environments. Thesis- Degree programme in Business Administration. Laurea University of Applied Sciences, Hyvinkää.

34. Rangnekar. A., 2004, Indian Leadership Traits - Back to the future, Narsee Monjee Institute of Management Studies, Mumbai.

35. Rockstuhl,T., Seiler,S., Ang,S., Dyne,L.V., and Annen,H., 2011., Beyond general intelligence (IQ) and emotional intelligence (EQ): The role of cultural intelligence (CQ) on cross-border leadership effectiveness in a globalized World. Journal of Social issues, Vol.67, No.4

36. Routamaa,V., and Debnath,N., 2011., Relationships between culture and values: A comparative study between Finland and India. University of Vaasa, Finland.

37. Routamaa,V., 2013., Cultural challenges to leadership: Work values in China, Finland and India. Proceedings of 8th Asian Business Research conference 1-2 April 2013, Bangkok, Thailand, ISBN: 978-1-922069-20-7.

38. Salyards,D.M., 2000., Doing Business in India: A Minnesota Entrepreneurs experiences. Family Business Review, Vol XIII,no4.

39. Sebastian.R., Parameswaran.A., and Yahya.F. (2006), Doing business in India. New Zealand Journal of Asian studies, 8(1) 17-40.

40. Schlosser,O., 2006., Humane orientation-A cross cultural study in 26 countries. Diploma Thesis. Department of Psychology, Justus-Liebig-Universitat Gieben, Germany.

41. Sinha,J.B.P., and Kumar,R., 2004., Methodology for understanding Indian culture. The Copenhagen Journal of Asian studies, Volume 19.

42. Singh,N., and Krishnan,V.R., 2007., Transformational Leadership in India: Developing and validating a new scale using grounded theory approach. International Journal of crosscultural management, Sage publications.

43. Struck,N.,2011.,The influence of the global leadership and organizational behavior effectiveness cultural dimensions on strategy execution: The example of German companies in India and Indian companies in Germany. Faculty of Business and Law. Aukland University of Technology

44. Taleghani,G., Salmani,D., and Taatian, A., 2010., Survey of leadership styles in different cultures. Iranian Journal of management studies, Vol.3.No.3.

45. Terlutter,R., Diehl,S., and Mueller,B.,2006., International advertising and communication. DUV Deuscher Universitats-Verlag, ISBN: 978-3-8350-0455-9.

46. Tenopir,C., 2008., The eagle and the elephant: Comparing organization culture in the United states and India. University of Tennessee, Knoxville.

47. Vesterinen,S., 2009., Leadership styles of Finnish nurse managers and factors influencing it. Journal of Nursing Management 17, 503-509.

7

EVALUATING STUDY FOR ELEMENTS AFFECTING TOURISM AWARENESS IN JORDAN: APPLIED ON AQABA AND WADI RUM AREAS

Dr. Farouk Abdelnabi Hassanein Attaallah1,
Noora Ibrahim Al-Ehewat2

1 Faculty of Tourism and Hotels Fayoum University – Fayoum, Egypt

2 Tourism Department- ASEZA Aqaba- Jordan

ABSTRACT

The objectives of current study are: to clarify issues relating to the factors affecting the tourism awareness, as well as stand on the current situation with regard to the concept of tourism awareness in the areas of study. The study indicated that the role of educational institutions and their contribution to increasing awareness of tourism confined in a positive direction. Moreover, the study cleared that there is a positive relationship between the work in the tourism sector and its contribution to increasing awareness of tourism in Aqaba and Wadi Rum Areas. It was found that there are significant differences between the mean of subgroups of respondent variable, as well as the presence of weak correction between the study areas and their impact on respondents answers. The study recommends holding awareness seminars and lectures to increase tourism awareness and publicize the importance of tourism, especially in non-urban areas, as well as the participation of the community to work to improve the level of tourism awareness they have.

KEYWORDS

Tourism Awareness, Aqaba, Wadi Rum, Educational Institutions, Jordan

1. INTRODUCTION

1.1 Background of Problem

The term "consciousness" took its semantic evolution towards processions to raise the intellectual and cultural life. It is significant to combine and save to signify an understanding and good perception, since consciousness is the result of mental and emotional complex coformed of thought, intuition, imagination , feelings , emotions , the will , conscience, the principles, values, foundations of instinct, life events and social systems and circumstances surrounding the man's life. This tremendous mixture of components awareness works very complex and each component contributes rate varies from one person to another, which endows each person, a kind of awareness is different from the consciousness of other (Bakkar, 2002, p: 1).

The tourism awareness is mainstay of growth and development of tourism activity anywhere. Purpose of allocating a day or a week or a year for tourism is to aware, to enlighten, and to glance the importance of tourism and its role in various fields. For that the World Tourism Organization allocated twenty-seventh of September of every year to be a day of global tourism under certain slogan selected by the World Tourism Organization to address an issue or to shed light on the different roles carried out by tourism (Jafra News, 2012, p: 1).

In Jordan, it is important that the tourism awareness links tourism education with the needs of the labor market, which represents one of the main pillars of the future of tourism in Jordan in both its internal and external; because young boys and girls representing the future demand for domestic tourism. Also they represent an attractive element for incoming international tourism through conscious awareness of Ethics for Tourism and culture, and through the involvement of many of them in career for the tourism sector. This may contribute to the increased polarization for those interested in tourism from outside Jordan, whether from the Gulf States, Arab or from around the world, On the other hand, there is need to focus on the national responsibility affiliated to system of education in Jordan to supply the tourism sector with qualified candidates to meet multiple needs in various fields related to the tourism sector (Bzazaw, 2012, p: 1).

1.2 Problem Statement

Shortage of tourism awareness among a broad range in the Jordanian society.

1.3 Hypotheses Of study

There is a positive and direct correlation between tourism education level and the level of tourism awareness among the local community.

- There is a positive and direct correlation between level of tourism awareness and work in the tourism sector.
- There is a positive and direct correlation between level of modernalization of the two regions and the level of awareness of tourism there.
- These hypotheses and the degree of correlation of coupling variables are tested at significance degree 5%.

1.4 Objectives of Study

- Stand on the level and type of education in the areas of study and its impact on tourism awareness.
- Stand on the types of work and jobs in the tourism sector and their impact on tourism awareness.
- Stand on the cultural level of the two regions.

1.5 Methodology of Study

- Descriptive analytical method.
- The case study technique.
- Simple random sample.
- Statistical program SPSS-V.17.

2. LITERATURE REVIEW

2.1 Tourism in Aqaba and Wadi Rum

The city of Aqaba is a strategic area and the only sea port of Jordan, and the city has many important industrial facilities, free trade zones, and King Hussein International Airport. It is an important center in South of Jordan and an origin of phosphate. Tourism in the city of Aqaba is active for various reasons; so it is coastal area located on the Red Sea, a nice place to visit, and an area filled with tourists who come to enjoy the wonderful sea and the nice atmosphere. More over in Aqaba the most important projects interested in eco-tourism in Jordan to observe birds, which attract tens of thousands and perhaps millions of migratory birds during the journey between Europe and Africa during the seasonal migration in the fall and spring. Furthermore, plant of Aqaba for bird watching includes the forest for the large trees and resident gardens trees in addition to large bodies of water. That all play in an integrated manner to attract different types of birds, some of which may be rare in the world, and which drives many interested to monitor birds and ornithologists to visit the area and to conduct of scientific analysis and research on the science of the birds. The Aqaba in particular and the region of Jordan in general are areas of global significance for bird migration based on the classification of International Birdlife. And embraces Marine Science Station located on the southern shore of the Gulf of Aqaba. Gallery of Marine Life where visitors can see the different types of coral, fish and other organisms endemic to the Gulf Aqaba. (www.Wikipedia.org, 2012).

Given the Wadi Rum, also called the Valley of the Moon due to the similarity of topography with lunar craters, and the advantage of the presence of mountains towering relatively there in, it is subject to the highest mountain peaks in southern Levant: the Mount or the bloody and Mount Ram, began Promoting this area began from the late eighties after making the movie "Lawrence of Arabia" in the sixties. Now tourism has become the source of income for many people who work as guides or any other business. Jordanian Ministry of Tourism markets Wadi Rum as a part of the tourism Golden Triangle, which includes Wadi Rum, Petra and Aqaba. Tourist activities include camping in this region and tours through the mountains on horseback and camels, or by using an SUV. And also include mountain climbing.(www.Wikipedia.org, 2012).

2.2 Importance of the tourism awareness

The importance of tourism awareness is summarized up in the following (Bzazaw, 2012, p: 1):

- Enhancement of the tourism sector and to encourage the accommodation of tourists in the tourist destination for longer.
- To encourage investment in the tourism sector.
- Development of human resources working in the tourism and hospitality industry.
- Advancement of the local community host for tourism.

2.3 Determinants of tourism awareness

Determinants of tourism awareness can be summarized in the following points (Dndraoy, 1994, p: 1):

First: the social factors that may be a real cause in the low tourist awareness among the citizens such as widespread unemployment among most members of the community, which may lead some to fraud on the tourist in a variety of ways and means.

Second: the social factors that may lead to the low level of awareness of some of the workers in the field of tourism as follows:

- The existence of problems among employees that may impede their performance.
- Awarding the tourist work branch offices in tourist areas and archaeological sites to individuals most of whom have medium qualifications and this is not commensurate with the nature of their role in tourism education and counseling.
- Lack of focus on training programs and courses designed to provide workers with scientific and practical experience in how to raise awareness and development of tourism among the citizens in general and among those of tourist areas in particular

2.4 Case studies on tourism awareness

Case Study 1

- **The Title of the study:** Trends of citizens in Wadi Musa towards tourism.
- **The conductor of study:** (Altayel, 2003, pp: 1, 20)
- **The objective of the study:** This study aimed to examine trends of citizens in Wadi Musa towards tourism and their extent affected by type and sector, income, education and age. -Results of the study: The findings of this study need to increase awareness of the importance of tourism to the local community in the areas of social and political by creating job opportunities for the unemployed in the area of tourism so as to improve their attitudes towards tourism, as well as the need to involve young people in tourism activities and the involvement of tourists with the local community in heritage activities

Case study 2

- **The Title of the study:** the position of the local community towards tourism, a case study Umm Qais.
- **The conductor of study:** (Chanaq and Otoum, 2000, pp: 17.1).

- **The objective of the study:** This study aimed to detect the positions of Jordanian citizens in the town of Umm Qais toward tourism and cultural dimensions social, political and economic.
- **The results of the study:** This study to focus on scientific studies of different aspects of tourism, as well as the establishment of training centers for local people to make them aware of the importance of tourism as a source of income and to encourage residents to invest in tourism projects.

Case Study 3

- **The Title of the study:** tourism awareness of the community and its role in promoting. - The conductor of study: (Alqarta and bani Isa 2011, pp.: 3, 1).
- **The objective of the study:** to stand on the role of tourism awareness in the promotions process, of the tourist area.
- **The results of the study:** This study concluded the need to raise the cultural level of the inhabitants of the host, as well as comprehensive awareness campaigns for the local community, cooperation and coordination between the Ministries of Education in the development of tourism awareness programs. The study benefited from previous studies in the preparation of a questionnaire study and compares the results with statistical analysis.

3. METHODOLOGY

In this section the student gives details about the method of research used and analysis of the information gathered for the current study.

3.1 Descriptive Analytical Method and Case Study Technique

Descriptive Analytical Approach handles the current phenomenon as it is in the real field, whether by a qualitative technique; case study technique. The principal purpose of the descriptive method is to evaluate things and conditions in their natural cases. Regarding the case study method, it has the crux importance in data collection and analysis in Descriptive Approach.

3.2 SPSS for Data Analysis

For techniques of analysis, statistics models that are suitable for measuring and analyzing attitudes and opinions towards a phenomenon are the Frequencies, the Mean, the Standard Deviation, the Standard Error of Mean, Test of Variance, ETA Value of Correlation, and ETA Square Value of Regression.

3.3 Sample and Questionnaire

The current research depends on the random simple sample for data collection.

The sample of the current study includes staff of tourism sector (6), tourist (5),Academics of the university of Jordan and Al-Balqa college(19), students of tourism and hospitality(8), and local community(37). Furthermore, this study adopted the Likert Scale, with choices strongly disagree = 1, disagree = 2, neutral = 3, agree = 4 and strongly agree = 5.

Seeing that questions, each answer has been given a special number by the number of answers in each question was as follows:

- Question No. (3): colleges and institutes tourism =1, tourist friends associations =2, training centers belonging to government agencies =3, training centers belonging to the tourism private sector =4, and other =5.

- Question No. (4): theoretical =1, practical =2, and theoretical and practical =3. - Question No. (6): academic position in tourism education institutions =1, job in government agencies of tourism =2, and jobs in the private tourism sector =3.
- Question No. (7): more awareness =1, less awareness =2, equal in the awareness =3, and I do not know =4.
- Question No. (8): Aqaba =1, Wadi Rum =2, and another region =3.

4. THE RESULT OF STUDY

We find that the most answers of respondents are in the positive direction, where the mean degree = 3.25, confined between the neutral and agree alternatives. There is dispersion between the respondents answers about the mean degree, where we find the degree of standard deviation = 1.357 greater than correct one; and there is no dispersion between the degree of the sample mean and the degree of the community mean where the value of the standard error of the mean was less than one = 0.157 .

In the continued table, we find that there are no significant differences between the means of sub-groups of respondent variable, where degree F = 0.133> degree of 0.05, as a result the null hypothesis is accepted; There is a weak correlation between the type of the study area and its impact on the answer of respondents, since the degree of ETA = 0.175, and the variable of the area of the study explains only (0.031) of the respondents answers.

We find that most of the answers of respondents are in the positive direction, where the mean degree = 3.81 confined between the neutral and agree alternatives. There is dispersion between the respondents answers about the mean degree , where we find the degree of standard deviation = 1.147 greater than correct one, and there is no dispersion between the degree of the sample mean and the degree of the community mean where the value of the standard error of the mean is less than one = 0.132 .

Table 1. Respondents answers about tourism awareness in Aqaba and Wadi rum.

Question	Frequencies						Mean	Std.Error of Mean	Std. Deviation
	S. Agree	Agree	Neutral	Disagree	S. Disagree	Total			
Educational institutions contribute to increase tourism awareness in the local community in Aqaba and Wadi Rum.	18.7%	34.7%	16%	14.7%	16%	100%	3.25	0.157	1.357

Table No.1 continued

Question	Significance F	Eta degree	Eta Squared
Educational institutions contribute to increase tourism awareness in the local community in Aqaba and Wadi Rum.	0.133	0.175	0.031

Table.2. Responses of respondents about the role of education and awareness-raising tour.

Question	Frequencies						Mean	Std.Error of Mean	Std. Deviation
	S. Agree	Agree	Neutral	Disagree	S. Disagree	Total			
Increased in the educational level led to increased tourist awareness.	34.6%	28%	22.7%	12%	2.7%	100%	3.81	0.132	1.147

Table No.2 continued

Question	Significance F	Eta degree	Eta Squared
Increased in the educational level led to increased tourist awareness.	0.118	0.182	0.033

In the continued table, we find that there are no significant differences between the means of sub-groups of respondent variable, where degree F = 0.118> degree of 0.05, According the null hypothesis is accepted; There is a weak correlation between the type of the study area and its impact on the answer of respondents, since the degree of ETA = 0.182, and the variable of the area of the study explains only (0.033) of the respondents' answers.

Table 3. Respondents answers about sources of education in the Wadi Rum and Aqaba

Question	Frequencies							Mean	Std Error of Mean	Std. Deviation
	Tourism faculties and college	Tourism societies	Governmental training centers	Private training centers	Others	More than one choice	Total			
The most important sources of tourism education in the areas of study (you can choose more than one source):	38.7%	12%	8%	5.3%	25.3%	10.7%	100 %	2.99	0.223	1.935

We find that the most answers of respondents are in the Tourist Friends Associations and training centers belonging to government agencies alternatives where the mean degree = 2.99.

There is dispersion between the respondents answers about the mean degree, where we find the degree of standard deviation = 1.935 greater than correct one, and there is no dispersion between the degree of the sample mean and the degree of the community mean where the value of the standard error of the mean is less than one correct = 0.223.

Table No.3 Continued

Question	Significance F	Eta degree	Eta Squared
The most important sources of tourism education in the areas of study (you can choose more than one source):	0.867	0.020	0

In the continued table, we find that there are no significant differences between the means of sub-groups of respondents variable, where degree F = 0.867> degree of 0.05, consequently the null hypothesis is accepted; There is a weak correlation between the type of the study area and its impact on the answer of respondents, since the degree of ETA = 0.020, and the variable of the area of the study explains (0) of the respondents answers.

Table 4. Responses respondents about the predominant state of education in the areas of study.

Question	Frequencies				Mean	Std.Error of Mean	Std. Deviation
	Theoretical	Practical	Theoretical and practical	Total			
Predominant status in tourism education	46.6%	10.7	42.7%	100%	2	0.112	0.973

We find that most of the answers of respondents are in the only practical alternative, where the mean degree = 2. There is no dispersion between respondents answers about the mean degree, where we find the degree of standard deviation = 0.973 less than correct one, and there is no dispersion between the degree of the sample mean and the degree of the community mean where the value of the standard error of the mean is less than = 0.112.

Table 4 Continued

Question	Significance F	Eta degree	Eta Squared
Predominant status in tourism education	0.210	0.146	0.021

In the continued table, we find that there are no significant differences between the means of sub-groups of respondent variable, where degree F = 0.210> degree of 0.05, so the null hypothesis is accepted; There is a weak correlation between the type of the study area and its impact on the answer of respondents, since the degree of ETA = 0.146, and the variable of the area of the study explains only (0.021) of the respondents answers.

Table .5. Responses of respondents about working in tourism jobs.

Question	Frequencies						Mean	Std.Error of Mean	Std. Deviation
	S. Agree	Agree	Neutral	Disagree	S. Disagree	Total			
Work in one of the jobs in the tourism sector increases the awareness of tourism.	30.7%	22.7%	24%	14.7%	8%	100%	3.53	0.149	1.288

We find that most of the answers of respondents are in the positive direction, where the mean degree = 3.53 confined between the neutral and agree alternatives. There is dispersion between the respondents answers about the mean degree , where we find the degree of standard deviation = 1.288 greater than correct one, and there is no dispersion between the degree of the sample mean and the degree of the community mean where the value of the standard error of the mean is less than one = 0.149 .

Table No.5 Continued

Question	Significance F	Eta degree	Eta Squared
Work in one of the jobs in the tourism sector increases the awareness of tourism.	0.314	0.118	0.014

In the continued table, we find that there are no significant differences between the means of sub-groups of respondent variable, where degree F = 0.314> degree 0.05, as a result the null hypothesis is accepted; There is a weak correlation between the type of the study area and its impact on the answer of respondents, since the degree of ETA = 0.118, and the variable of the area of the study explains only (0.014) of the respondents answer

Table 6. Responses of respondents about the jobs that since tourism awareness.

Question	Frequencies					Mean	Std.Error of Mean	Std. Deviation
	Academic job in tourism college	Formal government job	Work in private sector	More than one choice	Total			
From the point of view of you to the most important functions that achieve tourism awareness (you can choose more than one job).	29.3%	20%	36%	14.7%	100%	2.36	0.122	1.061

We find that most of the answers of respondents confined between the Academic position in tourism education institutions and position in the private tourism sector alternatives, where the mean degree = 2.36.There is dispersion between the respondents answers about the mean degree, where we find the degree of standard deviation = 1.061 greater than correct one, and there is no dispersion between the degree of the sample mean and the degree of the community mean where the value of the standard error of the mean is less than one = 0.122 .

Table No.6 Continued

Question	Significance F	Eta degree	Eta Squared
Work in one of the jobs in the tourism sector increases the awareness of tourism.	1	0	0

In the continued table, we find that there are no significant differences between the means of sub-groups of respondent variable, where degree F = 1> the degree of 0.05, so the null hypothesis is accepted; There is a weak correlation between the type of the study area and its impact on the answer of respondents, since the degree of ETA = 0, and the variable of the area of the study explains (0) of respondents Answers.

Table.7. Respondents answers compared to urban areas to other regions

Question	Frequencies					Mean	Std.Error of Mean	Std. Deviation
	More awareness	Equal awareness	Less awareness	I don't know	Total			
Urban areas compared with other regions	42.7%	33.3%	6.7%	17.3%	100 %	1.99	0.127	1.097

We find that the most answers of respondents are in the more awareness and equal awareness alternatives since the degree of Mean = 1.99. There is dispersion between the respondents answers about the mean degree, where we find the degree of standard deviation = 1.097 greater than correct one, and there is no dispersion between the degree of the sample mean and the degree of the community mean where the value of the standard error of the mean is less than one = 0.127.

Table No.7 Continued

Question	Significance F	Eta degree	Eta Squared
Urban areas compared with other regions	0.159	0.164	0.027

In the continued table, we find that there are no significant differences between the means of sub-groups of respondent variable, where degree F = 0.159> degree of 0.05, according the null hypothesis is accepted; There is a weak correlation between the type of the study area and its impact on the answer of respondents , since the degree of ETA = 0.164, and the variable of the area of the study explains only (0.027) of the respondents answers.

Table 8. Responses of respondents about their place of residence

Question	Frequencies				Mean	Std.Error of Mean	Std. Deviation
	Aqaba	Wadi rum	Other	Total			
You are from	46.7%	42.7	10.7	100%	1.64	0.077	0.671

We find that most of the answers respondents between the Aqaba and wadi Rum alternatives where the mean degree = 1.64. There is no dispersion between respondents answers about the mean degree, where we find the degree of standard deviation = 0.671 less than the correct one, and there is no dispersion between the degree of the sample mean and the degree of the community mean where the value of the standard error of the mean was less than one = 0.077.

Table No.8 Continued

Question	Significance F	Eta degree	Eta Squared
You are from	0	0.425	0.180

In the continued table, we find that there are significant differences between the means of sub-groups of respondent variable, where degree F = 0< degree 0.05, so the null hypothesis is rejected and we accept the alternative hypothesis, which means that there is significant difference between the means. There is a weak correlation between the type of the study area and its impact on the answer of respondents, since the degree of ETA = 0.425, and the variable of the area of the study explains only (0.180) of the respondents answers.

Table. 9 Responses of respondents about tourism activities.

Question	Frequencies						Mean	Std.Error of Mean	Std. Deviation
	S. Agree	Agree	Neutral	Disagree	S. Disagree	Total			
There are contribution by visitors to the area in the activities of the region	6.7%	40%	33.3%	10.7	9.3%	100%	3.24	0.121	1.051

We find that the most answers of respondents are in the positive direction, where the mean degree = 3.24 confined between the neutral and agree alternatives. There is dispersion between the respondents answers about the mean degree, where we find the degree of standard deviation = 1.051 greater than one correct, and there is no dispersion between the degree of the sample mean and the degree of the community mean where the value of the standard error of the mean is less than one = 0.121 .

In the continued table, we find that there are significant differences between the means of sub-groups of respondents variable, where degree F = 0.009< degree of 0.05, as a result the null hypothesis is rejected, and accept the alternative hypothesis saying that there are significant differences between the averages as there is a weak cancelation between the type of the study area and its impact on the answer respondents , since the degree of ETA = 0.298,and the variable of the area of the study explains only (0.089) of the respondents answers.

Table No.9 Continued

Question	Significance F	Eta degree	Eta Squared
There are contribution by visitors to the area in the activities of the region	0.009	0.298	0.089

5. CONCLUSION AND RECOMMENDATIONS

Study achieved its objectives, as clarified issues relating to the factors affecting the tourism awareness, as well as stand on the current situation with regard to the concept of tourism awareness in the areas of study

The study indicated that most of the answers about the role of educational institutions and their contribution to increasing awareness of tourism confined in a positive direction, therefore accepted hypothesis of the study, which indicate the presence of a positive relationship between the level of education and awareness in the areas of tourism study.

On the subject of work in one of the jobs in the tourism sector respondents in both regions painted out that there is a positive relationship between the work in the tourism sector and its contribution to increasing awareness of tourism, therefore accepted hypothesis that indicate the presence of a positive relationship between work in one of the sectors of tourism and increase awareness tourism.

Regarding zones study, we find that there are significant differences between the mean of sub-groups of respondent variable, as well as the presence of weak correction between the study areas and their impact on respondents Answers, therefore we rejected the hypothesis that there is a direct positive correlation between level of modularization of the two regions and the level of awareness, The For Answers investigator them about the existence of the involvement of the visitors to the two regions activities for those areas were the answers confined in the positive direction, which indicates the presence of active participation in certain activities.

The study recommends holding awareness seminars and lectures to increase tourism awareness and publicize the importance of tourism, especially in non-urban areas, as well as the participation of the community to work to improve the level of tourism awareness they have.

REFERENCES

1. Altayel, Hashem Mohamed (2003), Trends of Citizens in Wadi Musa towards Tourism, Mutah University, Karak, Jordan.

2. Alkarata, Nevein and Bani Isa, Ferial (2011), Tourism Awareness and its Role in the Promotion, Master Thesis, Amman, Jordan.

3. Bakkar, Abel-Kareim (2002), Renew of Awareness, Second Edition, Dar Alkalm, Damascus, Syria.

4. Dandrawy, Ali Abas (1994), Factors, Affecting Negatively the Development of Tourism Awareness with Perception of a Proposal for the Role of Social Service, Ph.D. Thesis, University of Fayoum, Egypt.

5. Chanaq, Mohamed and Otoum, Adnan (2000), Local Community Attitude towards Tourism, Case of Umm Qais, Irbid, Jordan.

6. http://www.jfranews.com, Accessed on October 2012. Bzazaw Ibrahim(2012), http://www.chihab.net, Accessed on October 2012. http://ar.wikipedia.org, Accessed on October 2012.

QUASI-FIXED INPUTS AND PERFORMANCE: THE CASE IN TAIWAN HOTEL INDUSTRY

Jui-Kuo Shang1, Fei-Ching Wang2, Wei-Ting Hung3,

1 Professor at the Department of Hotel Management, Jinwen University of Science and Technology, Taiwan.

2 Associate Professor at the Department of Hotel Management, Jinwen University of Science and Technology, Taiwan

3 Associate Professor at the Department of Business Administration, Vanung University, Taiwan

ABSTRACT

This study is the first to measure the efficiency when hotels employ quasi-fixed inputs with a two stage procedure. In the first stage, the dynamic data envelopment analysis developed by Nemoto and Goto (2003) is adopted to estimate the efficiency of 30 international tourist hotels in Taiwan between 2000 and 2012. In the second stage the bootstrapped truncatedregression proposed recently by Simar and Wilson (2007) is employed to analyze the operating characteristics may affect the performance of international tourist hotels. The empirical results indicate quasi-fixed inputs are almost the only source of overall inefficiency. From the bootstrapped truncated-regression analysis, this study indicated that hotels belonging to medium scale achieve better efficiency levels and chain operation is not the main determinant of the efficiency of international tourist hotels in Taiwan. These findings are not in line with most current studies and provide a good foundation and direction for future related research.

KEYWORDS

quasi-fixed inputs, data envelopment analysis, international tourist hotel, dynamic efficiency

1. INTRODUCTION

Economic prosperity and the emergence of the tourism industry are typical symbols of human progress. After the end of the Cold War, conflict and unrest were replaced by peace and harmony, which gradually reduced the obstacles to international travel. Furthermore, the rising trend of free trade resulted in burgeoning international exchanges, which increased mass travel by people from many countries. In addition to the rapid growth of the global travel industry, travel activities have also diversified. At the same time, travel has become a major socio-economic index in most parts of the world; since the tourism industry is a major foreign exchange earner for many countries, the sector promotes global economic development (Naisbitt, 1994, Gonzalez & Moral, 1996).

Recently emerging international tourism markets in the Asia Pacific include Taiwan, which was called the "Beautiful Island" by Portuguese mariners and is famous for its varied topography and diversified ecological resources. In order to mold a new image of Taiwan as an "island of tourism," the Tourism Bureau of the Ministry of Transportation and Communications has promoted the unique scenic and recreational attractions in Taiwan and has encouraged private participation in the construction of tourism and recreational facilities, including international tourist hotels (Weng and Wang 2004).

The number of international tourist hotels increased from 56 with a total of 17057 rooms in 2000 to 70 with a total of 20382 rooms in 2012. A survey by the Taiwan Tourism Bureau indicated that during the same period, each visitor to Taiwan in 2011 spent an average of US$293.21 per day. Of this total, spending in international tourist hotels made up US$116.58, making up 39.75% of the overall figure. Thus, the international tourist hotel industry is an essential component of the tourism industry in Taiwan.

The rapid growth of new international tourist hotels in Taiwan has intensified market competition. Thus, evaluating hotel efficiency is essential for hotel managers, not only for reference in decision-making, but also as the basis of any improvements. The importance of efficiency in the operations of the hotel industry is well supported and evidenced in the literature. However A review of current literatures about hotel efficiency indicates that most existing studies, to the best of our knowledge, have stayed within a static framework that cross sectional data is used to evaluate the efficiency of hotels. However, an analysis based on longitudinal data can provide more information about the performance trend. Furthermore, we do not anticipate that hotels will immediately be able to adjust their input levels to optimal levels, because some of the inputs, such as guest room, are quasi-fixed and thus cannot be adjusted to their optimal value in the short run. If we ignore this circumstance, the efficiency assessment results will be biased and give misleading conclusions.

By using a two-stage procedure, this study investigates the managerial efficiency of international tourist hotels in Taiwan. In the first stage, quasi-fixed inputs were applied to the dynamic data envelopment analysis (DEA) model developed by Nemoto and Goto (2003) to estimate the efficiency of 30 international tourist hotels during 2000-2012 by applying multiple operating data as both outputs and inputs. In the second-stage analysis, the bootstrapped truncated-regression model proposed recently by Simar and Wilson (2007) was used to analyze the operating characteristics that could not be quantified but might influence efficiency and might explain variation in managerial performance among hotels (Fried, Lovell and Schmidt 1993).

The rest of this study is organized as follows. Related literatures which have influenced this study are discussed in section 2. The third section provides a brief discussion of the methodology we used. The fourth section contains a discussion of data used in our study, and the empirical results are given in section 5. Conclusions are finally drawn in Section 6, along with recommendations for future research.

2. HOTEL EFFICIENCY LITERATURES

Performance evaluation is an essential aspect of management control. It not only provides a reference for decision-making, it also provides the basis of improvements. Therefore, measuring efficiency is an important and wide-ranging subject. It is commonplace to evaluate the efficiency of hotels on the basis of their inputs and outputs. Hotels are typically evaluated in terms of their ability to maximize output production with given inputs or to minimize input usage in the production of given outputs, relative to other hotels in some comparison set.

Early research on the hotel industry often used performance indicators. Most studies measured hotel performance by using single accounting and financial indicators in terms of ratios to measure hotel performance (Kims 1989; Wassenaar and Stafford 1991; Wijeysinghe 1993; Baker and Riley 1994; Huo 1994; Ens and Canina 2002)

Ratio analysis reveals the relationship between only two variables that can be evaluated at one time. However, this technique is clearly inadequate in the multi-output, multi-input settings that characterize real world organizations. In such a situation, where ratio analysis obtains a nebulous set of ratios that at best gives no clear indication of true efficiency and at worst conflicting indications (Sexton 1986). To avoid or to complement such complex arrays, one solution is synthesizing an index number as an aggregated index number. However, synthesizing such an index number requires a priori weighting to reflect the relative importance of its components to obtain a suitable efficiency measure. Such an effort could encounter numerous difficulties (Banker, Charnes, Cooper, Swarts and Thomas 1989).

In terms of frontier approach, two main techniques for sampling firms to construct an efficient production frontier are linear programming based (non-parametric approach) data envelopment analysis (DEA) and the regression-based (parametric approach) stochastic frontier approach (SFA). DEA was first introduced by Charnes, Cooper and Rhodes (1978, 1981) (CCR) and later extended by Banker, Charnes and Cooper (1984) (BCC), whereas SFA was first developed by Aiger, Lovell and Schmidt (1977) and Meeusen and Broeck (1977).

Anderson, Fish, Xia and Michello (1999), Barros (2004) and Chen (2007) employed SFA to assess the efficiency of hotel industry. The main advantages of SFA are its well-developed statistical tests of the validity of model specifications and its capacity to decompose the deviation from efficient levels between noise (or stochastic shocks) and pure inefficiency. However, the primary limitation of SFA is that researchers are required to specify an explicit, subjective, and in some cases, quite restrictive functional form. Another disadvantage is that researchers are required to make assumptions about the probability distribution of error terms. These arbitrary assumptions lead to different conclusions about the technical efficiency of a firm (Greene 1993).

In contrast, DEA is an efficiency assessment technique that does not require a priori weights or explicit specifications of a functional relationship. Nor does it provide an alternative framework for estimating the production frontier involving multiple inputs and outputs. Additionally, DEA also does not require a priori assignments of financial performance dimensions used in the evaluation process. It allows for the identification of appropriate benchmarks, and, above all, those decision making units (DMUs) which are performing poorly (Barros 2005). DEA is widely used to assess the efficiency of public and private sectors such as banks, airlines, hospitals, universities and manufacturers (Emrouznejad 2005). The measure has also been widely adopted to evaluate hotel efficiency in recent years.

Morey and Dittman (1995) was the first paper used DEA to measure the performance of hotel industry. It also provided a useful reference for input/output variables selection. The study revealed that among 54 hotel general managers, the average efficiency scores for managers were 89 percent. Following this line, Anderson, Fok and Scott (2000) applied DEA to evaluate the overall, allocative, technical, pure technical and scale efficiency of 48 United States hotels. The results revealed that the hotel industry was operating inefficiently with a mean overall efficiency score of 42%. The majority of efficiency losses are allocative in nature.

Tsaur (2000) employed DEA to estimate the efficiency of 53 international tourist hotels in Taiwan. The efficiencies of room departments and catering departments were calculated separately as 71% and 89%, respectively. Hwang and Chang (2003) also adopted the DEA model and the Malmquist productivity index to measure the managerial performance of 45 international tourist hotels in 1998 and the efficiency change of 45 international tourist hotels from 1994 to 1998 in Taiwan. The results revealed that there was a significant difference in efficiency change due to difference in source of customers and management style.

Chiang, Tsai and Wang (2004) used DEA to compare hotel performance between three ownership modes for international tourist hotels in Taiwan, independently owned and operated, franchise licensed, and managed by international hotel operator. The results showed that not all franchised or managed international tourist hotels performed more efficiently than the independent ones did. Barros (2005) estimated the efficiency of 43 Enatur hotels in Portugal by means of DEA. The results reveal that majority of the Enatur hotels are efficiency and the scale economies and location are major determinants of efficiency. Wang,

Hung and Shang (2006) used two-stage DEA to evaluate the relative cost efficiency of 49 international tourist hotels in Taiwan. The comparison results demonstrated that the international tourist hotel industry in Taiwan is inefficient and that most efficiency losses are attributable to technical inefficiencies, of which scale inefficiencies are the primary cause. Shang, Hung, Lo and Wang (2008) applied the three-stage DEA procedure to analyze the impact of ecommerce on the efficiency of hotels. After purging the effects of exogenous factors, this study found ecommerce adoption is not the main determinant of the efficiency of international tourist hotels in Taiwan. Based on the examples of 31 luxury and budget hotels in Korea, Min, Min and Joo (2009) illustrated the usefulness of DEA for the continuous improvement of hotel business practices. Cuiu, Ting and Huang (2010) employed system DEA to investigate the operating efficiency of Taiwanese hotels under different efficient frontier systems. They found if different frontier systems are assessed by the same frontier, then it will cause a wrong judgment in efficient reference sets and the efficiency of chain-operated hotels are higher than the efficiency of independent-operated hotels. Hsieh, Wang, Huang and Chen (2010) adopted the DEA model to assess the results of the relative efficiency and effectiveness of international tourist hotels in Taiwan to improve their operations. They also rank the hotels by applying VOKR method combined with entropy weight among the criteria. Chiu and Hung (2011) used DEA to measure the optimal occupancy rate, operational, and profitability efficiency of Taiwan's international tourist hotels. The efficiency results show that a high performance in operational efficiency does not necessarily ensure high profitability, increasing sales is not the best way to improve performance. Huang et al. (2012) employed Window-DEA and dynamic Tobit model to investigate the technical efficiency of hotel industry at the sub-national level in China. Ashrafi et al. (2013) applied non-radial DEA call the Slacks-Based Measure (SBM) to estimate the efficiency of hotel industry in Singapore from 1995 to 2010

Hotel efficiency is widely discussed in the literature, and the DEA approach is commonly used to explore this topic. However, some important points have not been considered. First, most of the abovementioned studies have stayed within a static framework, comparing the performance of hotels in the same period fail to model the intertemporal behaviour of hotels. Second, studies of hotel performance have reviewed and evaluated efficiency by assuming that input variables were variable inputs but have neglected the possible quasi-fixed property that hotels will not immediately be able to adjust their input levels to optimal values. Neglecting this circumstance, will bias the efficiency assessment results and lead to misleading conclusions. This study is the first to analyze longitudinal data for 2000 and 2012 to assess the dynamic efficiency of 30 international tourist hotels in Taiwan by considering quasi-fixed inputs

3. METHODOLOGY

According to the work of Nemoto and Goto (2003), the dynamic DEA model used to evaluate the efficiency of the Taiwan international tourist hotels can be set up as follows:

Suppose a firm use t x and t−1 k simultaneously in operating activities so as to generate t y and to hold t k at the end of the period where t x is a l ×1 variable inputs vector used in the period t , t k is a quasi-fixed input vector of m×1 at the end of the period t , and t y is an output vector of n×1 produced in the period t . Given N hotels with variable inputs, $X_t = (x_{t1}, x_{t2}, \ldots, x_{tN})$, quasi-fixed inputs at the beginning of the period t $K_{t-1} = (k_{t-11}, k_{t-12}, \ldots, k_{t-1N})$, quasi-fixed inputs at the end of the period t $K_{t-1} = (k_{t-11}, k_{t-12}, \ldots, k_{t-1N})$, and output $Y_t = (y_{t1}, y_{t2}, \ldots, y_{tN})$ the production possibility set Pt in period t is defined below:

$$P_t = \{(x_t, k_{t-1}, k_t, y_t) | X_t\lambda_t \le x_t, K_{t-1}\lambda_t \le k_{t-1}, \\ K_t\lambda_t \ge k_t, Y_t\lambda_t \ge y_t, \Sigma_{j=1}^{N}\lambda_{tj} = 1, \lambda_t \ge 0\}, \quad (1)$$

where λt is a N×1 intensity vector and λtj is the j th element.

Perfect anticipation is assumed for product demand and input price. The intertemporal efficient frontier of costs over the period from 1 to T are calculated as follows:

$$C = \min_{\{x_t, k_t\}_{t=1}^{T}} \{\sum_{t=1}^{T} \delta^t (w_t' x_t + v_t' k_{t-1} | (x_t, k_{t-1}, k_t, y_t)_{t=1}^{T} \in \times_{t=1}^{T} P_t, k_0 = \bar{k}_0\}, \quad (2)$$

where wt and t v are price of variable and quasi-fixed inputs in period t .δ is a constant discount factor. Overall efficiency (OE) from period 1 to T is calculated as the ratio of C to the corresponding actual costs.

$$OE = \frac{C}{\bar{C}}, \quad (3)$$

where C refers to the discounted sum of actual costs from period 1 to T .

The OE can be further decomposed into static efficiency (SE) and dynamic efficiency (DE). The SE represents inefficiencies generating from the levels of variable inputs while DE indicates inefficiencies originating from the paths of quasi-fixed inputs. The SE is represented by the efficiency costs given quasi-fixed inputs at observed levels:

$$C_{SE} = \min_{\{x_t\}_{t=1}^{T}}\{\sum_{t=1}^{T} \delta^t (w_t' x_t + v_t' \bar{k}_{t-1} | (x_t, k_{t-1}, k_t, y_t)_{t=1}^{T} \in \times_{t=1}^{T} P_t\}. \quad (4)$$

The difference between the actual costs C and CSE is result from the inefficient use of variable inputs since quasi-fixed inputs are held at observed levels. SE is measured by

$$SE = \frac{C_{SE}}{\bar{C}}. \quad (5)$$

Likewise, the gap between CSE and the efficient cost C results from choosing an inefficient path of quasi-fixed inputs. DE is then defined by

$$DE = \frac{C}{C_{SE}}. \quad (6)$$

Static efficiency (SE) can be further decomposed into static technical efficiency (TE) and static allocative efficiency (AE). The SE is represented by the ray minimum costs with quasifixed inputs as:

$$C_{TE} = \min_{\{\varphi_t\}_{t=1}^{T}}\{\sum_{t=1}^{T} \delta^t (\varphi_t w_t' \bar{x}_t + v_t' \bar{k}_{t-1} | (\varphi_t x_t, k_{t-1}, k_t, y_t)_{t=1}^{T} \in \times_{t=1}^{T} P_t\} \quad (7)$$

The radial contraction rate ϕ t is allowed to change over time. The TE is measured by

$$TE = \frac{C_{TE}}{\bar{C}} \quad (8)$$

In the frontier analysis framework, allocative efficiency (AE) is isolated by analyzing technical efficiency (TE) separately from cost efficiency. In this study, AE is defined as:

$$AE = \frac{C_{SE}}{C_{TE}} \tag{9}$$

AE indicates the costs that could have been preserved by the firms when the variable inputs were adjusted to optimal levels within the short-run period. From the equations (3), (5), (6), (8), (9), the following relationship can be obtained

$$OE = TE \times AE \times DE \tag{10}$$

In this study, the decomposition analysis based on (10) is used to measure the efficiency of international tourist hotels in Taiwan between 2000 and 2012.

4. DATA AND VARIABLES

4.1 Defining input-output variables

The data analyzed in this study were obtained from the Taiwan Tourism Bureau. The current administrative structure of the tourism industry in Taiwan is divided into the central and local government levels. At the central government level, the Tourism Bureau operates under the supervision of the Tourism Section of the Railways and Highways Department of the Ministry of Transportation and Communication. In Taipei and Kaohsiung Cities, which are directly administrated by the Executive Yuan, tourism affairs are handled by the Taipei Municipal Bureau of Transportation and Communications and by the Kaohsiung Municipal Bureau of Reconstruction respectively, and tourism sections operate under the construction bureaus of most other county and city governments of Taiwan Province. They conduct periodic nationwide surveys of the hotel industry. The latest published data set is 2012.

There are two types of hotels in Taiwan, international tourist hotels and ordinary tourist hotels. To improve the homogeneity of the research samples, the sample was limited to international tourist class hotels. After discarding hotels for which complete data were unavailable, the final data set consisted of 390 observations for 30 international tourist hotels during 2000-2012. Table 1 shows the basic information for these international tourist hotels.

With reference to the relevant research papers on domestic and foreign hotel performance and the business characteristics of the Taiwan hotel industry, this study selected output variables, input variables, input variable prices and quasi-fixed inputs as research variables. Outputs are defined as hotel revenues. The two main sources of revenue for international tourist hotels in Taiwan are accommodation and meals. These constitute over 80 per cent of overall hotel operating revenue. Other revenues include revenues from laundry operations, lease of store space, nightclubs, swimming pools, hairdressers, beauty salons, bookstores and service fees, which together constitute no more than 20 per cent of the total revenue. Similar to the output classifications employed by Hwang and Chang (2003), we divided hotel revenues into three broad categories: (1) room revenues (revenues from guest rooms); (2) F&B revenues (sales of food, snacks, alcohol, and beverages in dining rooms, coffee shops, banquets and night clubs) and (3) miscellaneous revenues (revenue from sources other than the room and F&B revenues).

The variable input was manpower. The numbers of full time employees were used to represent input manpower. Two input variables guest rooms (number of guest rooms in a hotel) and food and beverage (F&B) capacity (total floor area utilized by all such outlets in a hotel) were assumed as quasi-fixed factors. Based on the three selected input variables, input prices were defined as the average wage of full-time employees (P1), average room rate (P2) and the average cost of food and beverage operations (P3). The average cost of food and beverage operations equaled the ratio between total operating costs of food and beverage department and the allocated floor space area. Table 2 presents descriptive statistics for our data set. Input and output data are reported as the annual total and can be found in The Operating Report of International Tourist Hotel in Taiwan published by Taiwan Tourism Bureau. Table 3 shows the correlation matrix of outputs Oi and inputs i I . Note that all correlation coefficients are positive, which indicates that the inputs and outputs have an isotonic relationship. Therefore, these variables were included in the model (Golany and Roll 1989).

In the DEA model, the number of degrees of freedom will increase with the number of observations and decrease with the number of inputs and outputs. Thus Cooper, Seiford and Tone (2000) suggested that the number of observations should be at least maximum triple number of inputs plus outputs or the number of inputs multiply outputs. In this study the final data set consisted of 390 observations for 30 international tourist hotels during 2000-2011 which is larger than max $\{3\times3, 3(3+3)\}=18$.

4.2 Defining operating characteristic variables

Although the efficiency scores obtained from the dynamic DEA model represent the ability of international tourist hotels to convert inputs to outputs, it is possible that some differences in operating characteristics may affect the performance of international tourist hotels. Therefore, operating characteristics that cause variations in efficiency across international tourist hotels must also be analyzed

Location, management style and hotel size are considered the key operating characteristics that may affect hotel efficiency. There are two forms of location and management style. We construct dummy for location (1: city hotel; 0: resort hotel) and management style (1: chain operation; 0: independent) respectively. The dummy variable for hotel size is used to account for scale effect. The four hotel-size categories are small (less than 100 rooms), medium (100 to 300 rooms), large (300to 500 rooms), and very large (more than 500 rooms).

5. RESULTS AND DISCUSSIONS

This section summarizes the results obtained using the methods outlined in Section 3. Equation (10) was used to decompose the overall efficiency of Taiwan international tourist hotels. Table 4 shows the decomposition results for the dynamic DEA model. The second to fifth columns show the scores for overall efficiency and their decompositions. The OE scores ranged from 0.539 to 0.988, which indicated that, without any inefficiency, total cost reductions over the empirical period of were 0.12-46.1 percent. The decomposition of OE shows that most TE and AE are unity, so DE is very close to OE. This implies that little inefficiency is attributable to variable inputs and that the fixity of quasi-fixed inputs is the primary source of overall inefficiency. Put in another way, the efficiency level is improved if hotels are able to manage and adjust their quasi-fixed inputs efficiently. Therefore, ignoring the fixity of quasi-fixed inputs would bias the efficiency assessment results and obtain misleading conclusions.

A common practice in the DEA literatures, the Tobit regression analysis has been employed for analyzing which operating characteristics have an influence upon efficiency across international tourist hotels (Amemiya 1984, McCarty and Yaisawarng 1993, Kooreman 1994, Desli 1999), until Simar and Wilson (2007) has shown that such an approach would be inappropriate. Instead they justified a method based on the truncated-regression with bootstrap, and illustrated its satisfactory performance. In this study, we will adopt their method to analyze the relationship between efficiency scores and hotel operating characteristics that are not included within the dynamic DEA model.

Table 5 shows the bootstrapped truncated-regression results. The LOCATION variable was significantly and positively related to DE and OE, indicating that hotels located in city regions achieved better efficiency than those in resort areas. A plausible explanation is related to the cluster advantage emphasized by Porter (1998). He suggested that clusters increase the productivity of the companies based in the area and drive the direction and pace of innovation to support future productivity growth. Over 80 percent of the international tourist hotels In Taiwan are located in major metropolitan areas. Thus, the greater competition in metropolitan area has tended to increase efficiency levels. The other possible reason is that city hotels can not only attract leisure tourists, but possible some travelers that are doing business in downtown area. Thus, city hotels yield higher operating revenue and thus achieve higher efficiency score.

The idea of linking the variable of firm size to performance is traced back to Lall (1980). Current study such as Barros and Dieke (2008) showed that large African hotels are more efficient than small African hotels. However two studies on Taiwan international tourist hotels (Chen, 2007; Hwang and Chang, 2003) have found that there is no difference in efficiency between larger and small scale hotels. In the present study, we find the hotel scale variable (NEXT TO SMALLEST) was significantly and positively related to DE and OE. This means that in Taiwan the international tourist hotels with medium scale are more efficient than larger and small scale hotels. This result is not in line with all the mentioned studies. Erramilli and D'Souza (1993) claimed that medium scale enterprises often approach new investments carefully and minimize resource commitment in uncertain environment. Main (2001) further emphasized the major competitive advantage of medium scale hotels is the flexibility that they have compared to rigid standardized hotel brands, which permits customization and therefore, the ability to bring about niche markets. Thus, hotels belonging to medium scale are able to easily adjust their quasi-fixed input and reach batter dynamic efficiency level. It can also explain the phenomenon of fast-growing medium-sized business hotel industry in recent years in Taiwan.

With a rapid growth of international tourism markets in the Asia Pacific region, international tourist arrivals of Taiwan increase steadily. Because of this trend, more and more international tourist hotels were established, and would attempt to join international chains to get an international brand name and management expertise. Hotels that belong to international chains may be able to reduce their costs by sharing advertising, purchasing, management teams or reservation systems with fellow chain members, thus achieving economies of scale (Wang et al., 2006). However, hotels must weigh these merits against the costs of belonging to a chain. Hotels typically must pay a menu of fees for joining a chain and thus obtaining standardized operating processes and management knowledge. Additionally, hotels may find it extremely expensive to pass a stringent qualification evaluation and provide the services required to adhere organization quality standards (Mazzeo 2004). Therefore, the organizational mode of a hotel should choose for its expansion may depend on the costs and benefits associate with each form in its operating environment.

In terms of the relationship between efficiency and hotel management style (independent or chain operation), most current studies (Hwang and Chang, 2003; Wang et al., 2006; Chen, 2007; Barros and Dieke, 2008; Cuiu, Ting and Huang 2010) have found that the efficiency of chain operations is higher than that of independent operations. However results from this study are not in line with those studies. From Table 5, we can find that the MANAGEMENT STYLE variable is not significantly related to efficiency scores. In other words, the merits of chain affiliation may offset by the costs of becoming a chain member. Operating as part of a chain is not the main determinant of the efficiency of international tourist hotels in Taiwan.

6. CONCLUSIONS

Based on the longitudinal survey data obtained between 2000 and 2012, this study is the first research that employs dynamic DEA model and bootstrapped truncated-regression to analyze the dynamic efficiency of a sample of Taiwan international tourist hotels with quasifixed inputs.

The empirical results show that the international tourist hotel industry in Taiwan is inefficient: overall efficiency scores ranged from 0.539 to 0.988. This finding suggests that if hotels in our sample were required to perform as efficiently as best practice hotels total cost reductions over the empirical period would be 1.2-46.1 percent. Further analysis of the source of inefficiencies has led us to find that the fixity of quasi-fixed inputs is almost the only source of overall inefficiency. Therefore if international tourist hotels in Taiwan are able to manage and adjust their quasi-fixed inputs efficiently, the efficiency level will be improved.

From the Bootstrapped truncated-regression analysis, this study has found that hotels belonging to medium scale achieve better efficiency levels and chain operation is not the main determinant of the efficiency of international tourist hotels in Taiwan. These novel findings are different from most current studies.

Due to the limited available data, this study analyzed a sample of international tourist hotels in Taiwan. It should be cautioned that the findings of this study should not be overgeneralized. In fact, expanding the scope of this study would be worthwhile. Future studies may consider profit, quality of services, customer satisfaction, and quality of employees should be taken into account. Nevertheless, the findings provide a good foundation and direction for future related research.

Table 1. The Basic Information of 30 International Tourist Hotels in Taiwan

Hotel	Room scale	Management style	Location	Hotel	Room scale	Management style	Location
H01	4	1	1	H16	3	1	1
H02	3	1	1	H17	2	0	1
H03	3	0	1	H18	3	1	1
H04	1	0	1	H19	2	1	1
H05	2	0	1	H20	3	0	1
H06	3	1	1	H21	2	0	1
H07	2	0	1	H22	3	0	1
H08	2	0	1	H23	2	1	1
H09	2	1	1	H24	2	1	0
H10	2	0	1	H25	3	0	0
H11	4	1	1	H26	2	1	0
H12	4	1	1	H27	2	1	0
H13	4	1	1	H28	2	1	0
H14	4	1	1	H29	3	0	1
H15	3	1	1	H30	2	0	1

Note: Room scale (number of rooms): 1: smallest (less than 100 rooms), 2: next to smallest (100 to 300 rooms), 3: next to largest (300to 500 rooms), 4: and largest (above 500 rooms).
Management style: 0: independent, 1: chain; Location: 0: resort hotel, 1: city hotel.

Table 2. Descriptive Statistics of 30 Hotels in Taiwan (2000-2012)

Variable	Mean	Standard Deviation	Max.	Min.
Outputs				
O_1	261829856	247974554	1366496222	3113325
O_2	307936559	300242222	1301534840	5960371
O_3	102651566	140330598	596100145	255218
Inputs				
I_1	358	255	1069	59
I_2	333	165	873	97
I_3	1151	880	5298	48
Prices				
P_1	542157	148602	905858	130863
P_2	2836	1182	7401	1250
P_3	102359	53385	291687	21742

O_1 = Revenues from room department (NT$); O_2 = Revenues from food and beverage department (NT$); O_3 = Revenues other than O_1 and $O_2^{\cdot}$ (NT$); I_1 = Number of full-time employees; I_2 = Number of guest rooms; I_3 = Total area of meal department; P_1 = Average wage rate (NT$); P_2 = Average room rate (NT$); P_3 = Average production cost of meal department (NT$).

Table 3. Correlation Coefficients of Outputs and Inputs

	Output Variables			Input Variables		
	O_1	O_2	O_3	I_1	I_2	I_3
O_1	1					
O_2	0.51	1				
O_3	0.29	0.28	1			
I_1	0.58	0.27	0.29	1		
I_2	0.73	0.35	0.32	0.91	1	
I_3	0.33	0.68	0.49	0.82	0.55	1

Table 4. Efficiency Results for 30 Hotels in Taiwan

Hotel	OE	TE	AE	DE
H01	0.553	0.938	0.865	0.682
H02	0.978	1	1	0.978
H03	0.591	0.982	0.951	0.633
H04	0.787	1	1	0.787
H05	0.918	1	1	0.918
H06	0.744	0.989	0.993	0.758
H07	0.893	1	1	0.893
H08	0.923	1	1	0.923
H09	0.965	1	1	0.965
H10	0.792	0.940	0.925	0.911
H11	0.953	1	1	0.953
H12	0.988	1	1	0.988
H13	0.985	1	1	0.985
H14	0.937	1	1	0.937
H15	0.963	1	1	0.963
H16	0.926	1	1	0.926
H17	0.539	1	1	0.539
H18	0.863	1	1	0.863
H19	0.749	1	1	0.749
H20	0.628	1	1	0.628
H21	0.707	1	1	0.707
H22	0.609	1	0.919	0.663
H23	0.718	1	0.987	0.728
H24	0.898	1	1	0.898
H25	0.631	1	1	0.631
H26	0.638	1	0.969	0.658
H27	0.882	1	1	0.882
H28	0.958	1	1	0.958
H29	0.799	1	1	0.799
H30	0.907	1	1	0.907
Mean	0.815	0.995	0.987	0.827
Std. Dev.	0.136	0.012	0.028	0.123
Max.	0.988	1	1	0.988
Min.	0.539	0.938	0.865	0.539

Table 5. Results of Bootstrapped Truncated-regression

	OE		TE		AE		DE	
	Coefficient	*t*	Coefficient	*t*	Coefficient	*t*	Coefficient	*t*
INTERCEPT	0.2112	2.98***	0.1063	2.11**	0.1007	1.36	0.1809	3.91***
LOCATION	0.1989	3.82***	0.1751	1.61	0.1136	1.11	0.2021	3.68***
MANAGEMENT STYLE	0.3073	1.78	0.1336	1.37	0.1283	1.08	0.1501	1.49
SMALLEST	0.0823	0.97	0.1007	1.12	0.0987	0.68	1.0681	1.18
NEXT TO SMALLEST	0.1189	3.51***	0.2271	1.82	0.1032	1.39	0.1107	2.53**
NEXT TO LARGEST	0.0378	1.13	0.1537	1.26	0.0897	0.97	0.0782	1.66

Note: ***,**,* represents 1%, 5% and 10% level of significance respectively.

REFERENCES

1. Amemiya, T. (1984) Tobit models: a survey. Journal of Econometrics 24: 3-61.

2. Aigner, A., Lovell, C. and Schmidt, P. (1977) Formulation and estimation of stochastic frontier production function models, Journal of Econometrics, 86, 21-37.

3. Anderson R., Fish M., Xia Y., Michello E. (1999) Measuring efficiency in the hotel industry: A stochastic frontier approach, International Journal of hospitality Management, 18(1), 45- 57.

4. Anderson, R., Fok, R. and Scott, J. (2000) Hotel industry efficiency: An advanced linear programming examination, American Business Review, 18(1), 40-48.

5. Ashrafi, A., Seow, H. V., Lee, L. S., and Lee, C. G. (2013) The efficiency of the hotel industry in Singapore, Tourism Management, 37, 31-34.

6. Baker, M. and Riley, M. (1994) New perspectives on productivity in hotels: Some advances and new directions, International Journal of Hospitality Management, 13(4), 297-311.

7. Banker, R., Charnes, A. and Cooper, W. (1984) Some models for estimating technical and scale inefficiencies in data envelopment analysis, Management Science, 30 (9), 1078-1092.

8. Banker, R., Charnes, A., Cooper, W., Swarts, J. and Thomas, D. (1989) An introduction to data envelopment analysis with some of its models and their uses, Research in Government and Nonprofit Accounting, 5, 125-163.

9. Barros, C. (2004) A Stochastic cost frontier in the Portuguese hotel industry, Tourism Economics, 10, 177-192.

10. Barros, C. (2005) Measuring efficiency in the hotel sector, Annals of Tourism Research, 32(2), 456-477.

11. Barros, C. and Dieke, P. (2008) Technical efficiency of African hotels, International Journal of Hospitality Management, 27, 438-447.

12. Charnes, A., Cooper, W. and Rhodes, E. (1978) Measuring the efficiency of decision making units, European Journal of Operational Research, 2(6), 429-444.

13. Charnes, A., Cooper, W. and Rhodes, E. (1981) Evaluating program and managerial efficiency: An application of data envelopment analysis to program follow through, Management Science, 27(6), 668-697.

14. Chen, C. (2007) Applying the stochastic frontier approach to measure hotel managerial efficiency in Taiwan, Tourism Management, 28(3), 696-702.

15. Chiang, W., Tsai, H. and Wang, L. (2004) A DEA evaluation of Taipei hotels, Annals of Tourism Research, 31(3), 712-715.

16. Chiu, Y., Ting, C. and Huang, C. (2010) The different systems for tourist hotels efficiency estimation in Taiwan, Service Industries Journal, 30(14), 2367-2385.

17. Chiu, Y. and Huang, C. (2011) Evaluating the optimal occupancy rate, operational efficiency, and profitability efficiency of Taiwan's international tourist hotels, Service Industries Journal, 31(13), 2145-2162.

18. Desli, E. (1999) Estimation of technical efficiency from parametric and non-parametric production frontiers, PhD Dissertation in University of Connecticut. Emrouznejad, A. (2005) Measurement efficiency and productivity in SAS/OR, Computers and Operations Research,32(7), 1665-1683.

19. Enz, C. and Canian, L. (2002) Best of time, the worst of times: Difference in hotel performance following 9/11, Cornell Hotel and Restaurant Administration Quarterly, 43, 22- 32.

20. Erramilli, K. M. and D'Souza, D. (1993) Venturing into foreign markets: The case of small service firm, Entrepreneurship Theory and Practice, 17 (4), 29-41.

21. Fried, H., Lovell, C. and Schmidt, S. (1993) The measurement of productivity efficiency: Techniques and applications, Oxford University Press: Oxford.

22. Golany, B. and Roll, Y. (1989) An application procedure for DEA, Omega, 17, 237-250.

23. Greene, W. (1993) The econometric approach to efficiency analysis, in Fried, H., Lovell, C. and Schmidt, S. (ed.). The measurement of productive efficiency: Techniques and applications. Oxford University Press: Oxford.

24. Hsieh, L., Wang, L., Huang, Y. and Chen, A. (2010) An efficiency and effectiveness model for international tourist hotels in Taiwan, Service Industries Journal, 30(13), 2183-2199.

25. Huang, Y., Mesak, H. I., Hsu, M.k., and Qu, H. (2012) Dynamic efficiency assessment of the Chinese hotel industry, Journal of Business Research, 65(1), 59-67.

26. Hwang, S. and Chang, T. (2003) Using data envelopment analysis to measure hotel managerial efficiency change in Taiwan, Tourism Management, 24, 357-369.

27. Huo, Y. (1994) Internal environment, organizational form, and the impact on financial performance of hotel chains. PhD Dissertation in Virginia Polytechnic Institute and State University.

28. Kims, S. (1989) The basics of yield management, Cornell Hotel and Restaurant Administration Quarterly, 30, 14-19.

29. Kooreman, P. (1994) Nursing home care in the Netherlands: A non-parametric efficiency analysis, Journal of Health Economics, 13, 301-316.

30. Lall, S. (1980) Monopolistic advantages and foreign involvement by U. S manufacturing industry, Oxford Economic Papers, 32, 102-122.

31. Main, H. (2001) The use of the Internet by hotels in Wales – A longitudinal study: 1994- 2000, International Journal of Hospitality Information Technology, 2(2), 35-44.

32. Mazzeo, M. (2004) Retail contracting and organizational form: alternatives to chain affiliation in the motel industry, Journal of Economics and Management Strategy,13(4), 599-615.

33. McCarty, T. and Yaisawarng, S. (1993) Technical efficiency in New Jersey school districts, in Fried, H., Lovell, C. and Schmidt, S. (eds.). The Measurement of Productive Efficiency: Techniques and Applications. Oxford University Press: New York.

34. Meeusen, W. and Broeck, J. (1977) Efficiency estimation from Cobb-Douglas production functions with composite error, International Economic Review, 18(2), 435-444.

35. Min, H., Min, H. and Joo, S. (2009) A data envelopment analysis on assessing the competitiveness of Korean hotels, Service Industries Journal, 29(3), 367-285.

36. Morey, R. and Dittman, D. (1995) Evaluating a hotel GM's performance: A case study in benchmarking, Cornell Hotel Restaurant and Administration Quarterly, 36(5), 30-35.

37. Nemoto, J. and Goto, M. (2003) Measurement of dynamic efficiency in production: An application of data envelopment analysis to Japanese electric utilities, Journal of Productivity Analysis, 19, 191-210.

38. Porter, M. (1998) Clusters and the new economics of competition, Harvard Business Review, 76(6), 77-90.

39. Sexton, T. (1986) The methodology of data envelopment analysis, in Silkman RH. (ed.). Measuring efficiency: An assessment of data envelopment analysis. American Evaluation Association: San Francisco.

40. Shang, J., Hung, W., Lo, C. and Wang, F. (2008) Ecommerce and hotel performance: threestage DEA analysis, Service Industries Journal, 28(4), 529-540.

41. Simar, L. and Wilson, P. W. (2007) Estimation and Inference in Two-Stage, Semi-Parametric Models of Production Processes, Journal of Econometrics, 136, 1-34.

42. Tsaur, S. (2000) The operating efficiency of international tourist hotels in Taiwan, Asia Pacific Journal of Tourism Research, 6 (1), 29-37.

43. Wang F., Hung, W. and Shang, J. (2006) Measuring the cost efficiency of international tourist hotels in Taiwan, Tourism Economics, 12(1), 65-85.

44. Wassenaar, D. and Stafford, E. (1991) The lodging index: An economic indicator for the hotel/motel industry, Journal of Travel Research, 30(1), 18-21.

45. Wijeysinghe, B. (1993) Breakeven occupancy for a hotel operation, Management Accounting, 7(12), 32-33.

9

A COMPARATIVE STUDY ON THE IMAGES OF MALAYSIA AND THAILAND AS PERCEIVED BY JAPANESE UNDERGRADUATE STUDENTS

Bình Nghiêm-Phú

Keio University, Graduate School of Media and Governance, Kanagawa, Japan.

ABSTRACT

This paper reports the findings of a study on the images of Malaysia and Thailand. The study investigated and compared the perceptions of Japanese youth tourists of the two countries as potential tourism destinations. It reexamined the structure of destination image, and the relationship between the three components of destination image and the intention to visit each country of the potential tourists through a structural equation model. The findings of this study revealed the important role of emotional/affective perceptions in the formation of destination image and intention-to-visit, and generate implications for the projection and communication of the images of Malaysia and Thailand in the Japanese youth tourist market

KEYWORDS

destination image, intention-to-visit, Japanese undergraduate students, Malaysia, Thailand

1. INTRODUCTION

Destination image is a popular topic in tourism studies. The image of a destination can be compared to two sides of a coin. On the one side, it is the image perceived by tourists, residents, and other stakeholders. On the other side, it is the image created by destination managers/ marketers to attract more tourists, laborers, and investors. Destination image is not a static concept but develops over time. Consequently, even after four decades, research into destination image is still progressing.

Within a global framework, the phenomenon of "regional tourism" emerged in the 1990s (Chang 2004). One of the examples is Southeast Asia. This region is home to eleven countries (i.e., Brunei, Cambodia, Indonesia, Laos, Malaysia, Myanmar, the Philippines, Singapore, Timor-Leste, Thailand, and Vietnam), and is rich in natural and cultural resources due to the diversity of climate and ethnology. Southeast Asia is a well-known destination in the world tourism map. Among the member countries, Vietnam, Laos, Cambodia, Myanmar, TimorLeste, and Brunei are considered as "emerging" destinations, while the others might be referred to as "mature" ones (Richter 2009).

Many attempts have been made to investigate the image of each Southeast Asian country as perceived by the international tourists (e.g., Chena, Chen & Okumus 2013; Chens, Sok & Sok 2008; Litvin & Ling 2001; de Guzman et al. 2012; Talib 2005; Kamil 2010; Mohamad, Abdullah & Mokhlis 2012; Pawitra & Tan 2003; Yeung, Wong & Ko 2004; Henderson 2007; Rittichainuwat, Qu & Mongknonvanit 2002, 2008; Howard 2009; Rittichainuwat & Chakraborty 2009; Truong 2005; Truong & King 2006; Bui 2011). However, comparative studies to examine how the tourists from one market perceive the images of several countries have not yet been undertaken. This approach forms the first objective of this study. The study aims to investigate the images of Malaysia and Thailand as perceived by Japanese undergraduate students. Malaysia and Thailand were selected because they are the two largest tourism destinations in the region. (In 2013, Thailand had a new record of inbound tourist arrivals – approximately 26.7 million, whilst in the same year Malaysia hosted only approximately 1 million fewer arrivals.) The Japanese market was targeted because of its huge spending power and geographical proximity. Although in recent years, Japan has lost its leading position to China, this market's potential is still attractive to the outbound tourism destinations (ITB Berlin 2013). Due to economic reasons, Japanese travelers may cut down their spending on international holidays. In addition, the current relationships between Japan and China/Korea may deter their decisions to visit these countries. Furthermore, long-haul destinations to Europe, America, and Australia become more expensive and time-consuming. Consequently, traditional short-haul destinations, especially those with beaches, such as Taiwan, Hawaii, Saipan, and the Southeast Asian countries have the advantage. In 2013, Japan was the fourth largest generating market in Thailand (approximately 1,538,000 arrivals), and the ninth in Malaysia (approximately 513,000 arrivals). Within the Japanese market, the undergraduate students were targeted because as a sub-group of the youth tourist market, they are not much affected by the external conditions (UNWTO 2008). Moreover, these tourists possess the potential to become loyal tourists (repeaters) if they are satisfied with the destinations although they are often ignored by the governments around the world (UNWTO 2008). In the context of this study, Malaysia and Thailand are considered as potential tourism destinations, and the students' perceived images as pre-visit images.

In addition to the examination of the images of Malaysia and Thailand as perceived by Japanese undergraduate students, this study attempts to reconfirm the structure of destination image as tested by other researchers in the other settings outside of the Southeast Asia region (e.g., Baloglu 1996; Kim & Yoon 2003; Lin, Morais, Kerstetter & Hou 2007; San Martin & del Bosque 2008). This model proposes that, perceived destination image consists of three associated components: cognitive, affective, and overall. Moreover, the relationship between perceived destination image and potential tourists' intention-to-visit is also taken into account. In previous studies, it was found that intention-to-visit is under the influence of perceived destination image (e.g., Chiang, Lu & Chang 2003; Lee, Scott & Kim 2008; Nicoletta & Servidio, 2012).

This is an exploratory effort to apply the comparative approach to perceived destination image studies in the Southeast Asia region, which has largely been ignored by previous researchers. The findings will provide practical implications for the management and communication of each country's images.

2. LITERATURE REVIEW

2.1 Perceived destination image structure and its role as an antecedent of intention-tovisit

Perceived destination image is structured by two major components: cognitive, and affective. The cognitive element is the knowledge, perceptions, and beliefs that a perceiver holds about a destination; whereas the affective element reflects the feelings that he/she has toward the destination (Stepchenkova & Mills 2010). In addition to these two components, many researchers include another element into the model of destination image structure: the overall image (e.g., Baloglu 1996; Baloglu & McCleary 1999a; Kim & Yoon 2003; Lin, Morais, Kerstetter & Hou 2007; Molina, Frias-Jamilena & Castaneda-Garcia 2013). In these studies, researchers discovered that the three components of destination image have significant correlations. The affective element of destination image is significantly affected by the cognitive component, and the overall component is under significant influences of both the cognitive and affective ones. For potential tourists, the effect that affective image has on overall image compared to that of cognitive image is larger (Baloglu & McCleary 1999a; Kim & Yoon 2003; Lin, Morais, Kerstetter & Hou, 2007). With actual tourists, the reversed effect is observed. For example, in San Martin and del Bosque (2008), cognitive image factors were found to have higher loadings on destination image compared to that of affective image

Perceived destination image plays an important role in the destination selection process of potential tourists. Destination image can be considered as the "pull" component of tourists' motivation (Crompton 1979; Weaver & Lawton 2002). For the potential or first-time visitors, a destination is an intangible product that cannot be directly experienced before the actual consumption; therefore, they rely on the perceived images in deciding to choose one destination over another (Weaver & Lawton 2002). The final destination is selected from a group of potential alternatives (e.g., Crompton 1992; Kwon 2002; Page 2003). However, a destination need not have a good range of attractions and amenities; instead, its attributes must be favored by the potential tourists (UNWTO 2007). Besides the perceived images, the selection process is also affected by other internal and external factors, including sociodemographic characteristics, cultural background, motivations, interests, emotion, self-image, and attitudes (Chen 1996; Lumsdon 1997; Seddighi, Nuttall & Theocharous 2001; Chiang, Lu & Chang 2003; Stepchenkova & Mills 2010; Nicoletta & Servidio, 2012).

Previous studies have also examined the images of different destinations as perceived by tourists from a given market (e.g., Baloglu & McCleary 1999b; Reisinger & Turner 2000; Bozbay & Ozen 2008). The findings of these comparative studies reveal the destination which is more preferred by the tourists. It also provides the stakeholders on the supply side with the important information to project positive attributes to communicate to the potential tourists, and to guarantee high quality experiences to satisfy the actual tourists' expectations. In particular, these efforts will help create positive word-of-mouth, which is an important source of organic information utilized in the formation of destination image (Gartner 1993).

2.2 Japanese tourists' perceptions of Malaysia and Thailand

In Japan, Southeast Asia is usually classified as "Asia," with beach resorts such as Bali in Indonesia, Phuket in Thailand, Penang and Langkawi in Malaysia, and Cebu in the Philippines being described as "paradises" (Yamashita 2009). One of the reasons for Japanese tourists to choose this region is "nostalgia," rather than "exoticism." They prefer "sea," as compared to the preferred attraction of the European tourists to be "forests" (Yamashita 2009). However, there are few studies which directly investigate the images of the region and its member countries as perceived by Japanese tourists.

Tan, McCahon and Miller (2002) studied the economic factors that affected the travel decisions of international tourists to Indonesia and Malaysia. They found that for both destinations, income was the most important factor in case of the tourist flow from Japan. In another study, Wang and Wu (2003) measured the factors that influenced Japanese tourists to Hong Kong, Taiwan, Singapore, Malaysia, the Philippines, and Indonesia. The findings revealed that variation in exchange rates affected Japanese visitors to Taiwan and the Philippines adversely, but did not significantly affect Japanese visitors to Hong Kong, Singapore, Malaysia, and Indonesia. In addition, Wang and Wu (2003) found that "sightseeing" and "business" are two of the major purposes of Japanese travelers to visit the Southeast Asian destinations. Particularly in the case of Malaysia, Japanese travelers prefer the isolated, unpolluted beach/sea, and the coastal forest landscape images of the country (Talib 2005).

In the case of Thailand, Sangpikul (2008) examined the motivations of Japanese senior travelers to visit the country. The motivational factors included "novelty and knowledgeseeking," "rest and relaxation," and "ego-enhancement." In addition, these tourists were attracted by the "cultural and historical attractions," "travel arrangements and facilities," "shopping and leisure activities," and "safety and cleanliness" of Thailand.

Given the importance of the Japanese market and tourism development in Malaysia and Thailand, the number of studies which directly investigate the images of the two countries as perceived by Japanese tourists is too limited. The existing literature cannot provide enough evidences to directly compare and distinguish Malaysia and Thailand's images.

3. STUDY OBJECTIVES AND OPERATIONALIZATION

This study aims to measure the images of two Southeast Asian destinations – Malaysia and Thailand – as perceived by Japanese undergraduate students. Specifically, its objectives are: (1) to compare the perceived images of Malaysia and Thailand (Figure 1), and (2) to confirm a theoretical model regarding the structure of perceived destination image and its effects on potential tourists' intention to visit a destination (Figure 2).

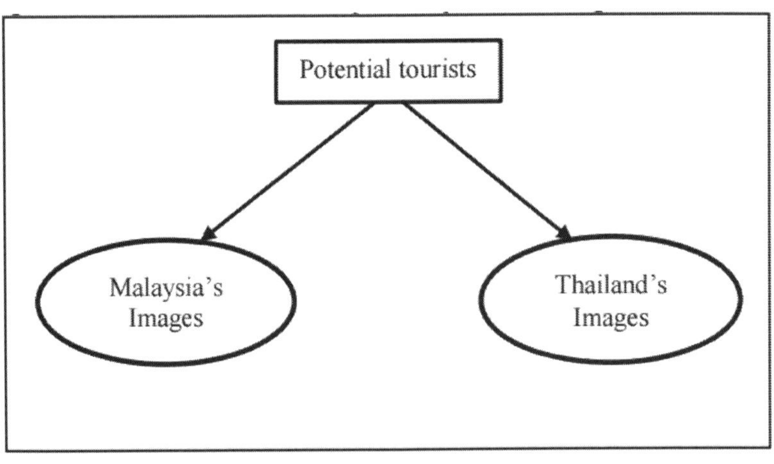

Figure1. Theoretical model to compare the perceived images of two countries

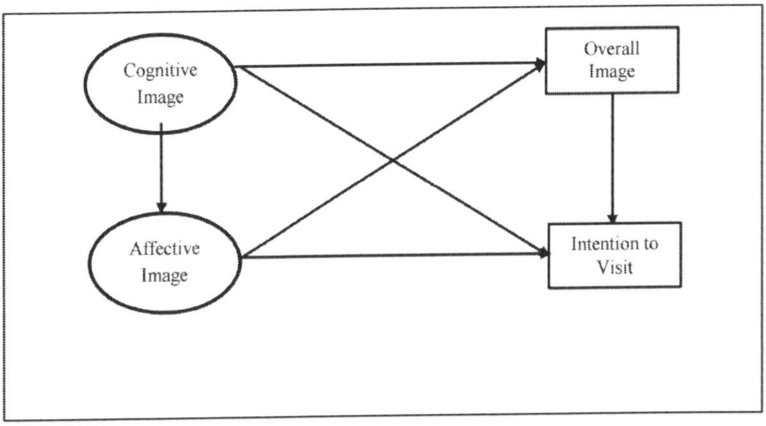

Figure 2. Theoretical model assuming the associations among the three components of perceived destination image and intention-to-visit.

Within the second objective, two hypotheses are tested in a replication of the findings of previous studies (e.g., Baloglu 1996; Kim & Yoon 2003; Lin, Morais, Kerstetter & Hou 2007; San Martin & del Bosque 2008; Chiang, Lu & Chang 2003; Lee, Scott & Kim 2008; Nicoletta & Servidio, 2012; Molina, Frias-Jamilena & Castaneda-Garcia 2013). It is assumed that: H1. Perceived destination image is structured by three correlated components, with (a) cognitive image significantly affects affective image, (b) cognitive image significantly affects overall image, and (c) affective image significantly affects overall image. H2. Perceived (a) cognitive image, (b) affective image, and (c) overall image significantly influence potential tourists' intention-to-visit.

In order to compare the images of the two countries, and to confirm the theoretical model, this study applied a quantitative approach. A self-administered questionnaire was utilized as the study instrument. The measures of this study were collected from the literature. Fourteen items to measure the cognitive image were generated from a combination of English, Japanese, and bilingual documents (Baloglu 1996; Miyamori 2001; Lee 2002; JTB Corp. 2010). Four bipolar items to measure the affective image were adopted from Russell and Pratt (1980). The application of these attributes is popular within the existing literature, and they are able to capture the most important and common images of places (see, for example, Echtner & Ritchie 2003). There were one item to capture the overall perception, and one item to measure the intention to visit the destinations in the next five years of the respondents. Following previous studies on Japanese tourists (e.g.,

Andersen, Prentice & Watanabe 2000; Sangpikul 2008), a five-point Likert-type scale was used in the instrument. The profile of the respondents included the information regarding their sex, international experiences, school year, and part-time experiences.

After generation, all the items written in English were translated into Japanese by the researcher. The translation was then discussed with the Japanese students in a graduate school in tourism in Saitama prefecture, Japan to guarantee its appropriateness and smoothness. The Japanese version was approved by two professors who are fluent in both Japanese and English. It was then pre-tested on a group of 50 students; small adjustments were made after that to improve its clarity.

The main survey was conducted in July, 2012 with undergraduate students in three campuses of two universities in Tokyo, Saitama, and Chiba prefectures (in the East side of Japan). Considering the constraints of time and resources, a nonprobability sampling method was adopted. The respondents were approached conveniently, and 357 students agreed to answer the questionnaires. However, 46 questionnaires which did not contain all the information regarding the images of the two countries were eliminated; thus, the usable questionnaires were 311. The data were analyzed using SPSS 16 and AMOS 21. The findings of this study are presented in the following section.

4. FINDINGS AND DISCUSSION

Table 1 describes the profile of the respondents. There were 128 male and 183 female students; those aged 18-20 accounted for the majority of the sample. More than half of the sample was composed of first year students (158). A total of 194 students stated that they had overseas travel experiences, and most of them had part-time job experiences (232). Among them, only twelve have visited Malaysia, and sixteen have been to Thailand.

4.1 The comparison of the images of Malaysia and Thailand

The image attributes of Malaysia were evaluated slightly higher than these of Thailand with the exceptions of the overall image and five cognitive items: C1, C6, C7, C10, and C11 (Table 2). Among the image items, only Thai food got a mean score of 4 points. Malaysia had more advantages with their nature, nature-related attractions, infrastructure, and safety. The advantages of Thailand include their cultural attractions. Several attributes of both countries received less than 3 points of evaluation: C4, C5, C9, and C13.

Table 1. Profile of the respondents.

		n	%			n	%
Age	18	92	29.58	Year in university	1	158	50.80
	19	99	31.83		2	105	33.76
	20	56	18.01		3	45	14.47
	21	20	6.43		4	3	0.96
Sex	Male	128	41.16	Overseas travel experience	No	115	36.98
	Female	183	58.84		Yes	194	62.38
				Part-time job experience	No	76	24.44
					Yes	232	74.60

To compare the perceptions of Japanese undergraduate students of the two countries, *t*- test was computed in SPSS (Table 2). The perceptions of these potential tourists were significantly different in nine cognitive image attributes, including C2, C3, C4, C5, C7, C8, C11, C12, and C13. Their evaluations did not differ significantly in case of the affective attributes, and the overall perception. This result is quite consistent with the observation made by Yamashita (2009). Yamashita (2009) stated that Southeast Asia is classified by Japanese tourists as one destination; in other words, their member countries' images may be perceived similarly in this market. Although not significantly differed, Japanese potential youth tourists' intention to visit Thailand was stronger than that to Malaysia.

4.2 The confirmation of destination image structure, and relationship between destination image and intention to visit

Analyses with structural equation modeling (SEM) in AMOS were conducted to confirm the structure of destination image and other relationships. The minimum criteria to accept the fit indices of a model are as follows: $2 < \chi 2 /df < 3$, $0.05 < RMR < 0.10$, $0.05 < RMSEA < 0.08$,

0.90 < NFI < 0.95, 0.95 < CFI < 0.97, 0.90 < GFI < 0.95, and 0.85 < AGFI < 0.90, with χ^2 = Chi square, df = degree of freedom, SRMR = Standardized Root Mean squared Residual, RMSEA = Root Mean Square Error of Approximation, NFI = Normed Fit Index, CFI = Comparative Fit Index, GFI = Goodness-of-Fit Index, and AGFI = Adjusted Goodness-of-FitIndex (Schermelleh-Engel, Moosbrugger & Müller 2003). Chi square and its significance level are not considered as a fit index because Chi square is too sensitive (Schermelleh-Engel, Moosbrugger & Müller 2003). Two solutions to verify the theoretical model illustrated in Figure 2 were applied.

Table 2. Comparing respondents' perceptions of and intentions to visit Malaysia and Thailand

Code	Attribute	Malaysia	Thailand	*p*
	Cognitive images			
C1	Good value for money	3.42	3.52	.146
C2	Beautiful scenery/natural attractions	*3.68*	*3.40*	*.000*
C3	Beautiful beaches	*3.46*	*3.14*	*.000*
C4	Good climate	*2.94*	*2.80*	*.034*
C5	Standard hygiene and cleanliness	*2.79*	*2.59*	*.005*
C6	Famous cultural attractions (art galleries, museums, etc.)	3.14	3.23	.206
C7	Famous historical attractions (historic sites, architecture, etc.)	*3.31*	*3.46*	*.036*
C8	Marine sports (wind surfing, scuba diving, etc.)	*3.19*	*2.95*	*.002*
C9	Entertainment (concerts, music, etc.)	2.86	2.86	.963
C10	There are many things to shop	3.25	3.28	.667
C11	Appealing local cuisines	*3.68*	*4.02*	*.000*
C12	Quality of infrastructure	*3.10*	*2.94*	*.022*
C13	Personal safety	*2.84*	*2.67*	*.010*
C14	Friendly people	3.38	3.37	.931
	Affective images			
A1	Unpleasant-Pleasant	3.56	3.45	.100
A2	Sleepy-Arousing	3.71	3.80	.234
A3	Distressing-Relaxing	3.31	3.16	.060
A4	Gloomy-Exciting	3.68	3.65	.663
	Overall perception and Intention-to-visit			
O	Overall perception	3.60	3.61	.882
I	Intention-to-visit	3.12	3.25	.136

Note. Significant differences are highlighted in *italic*.

4.2.1 First analysis solution

In the first solution, factor analysis was firstly conducted to clarify the latent structure of the cognitive and the affective images of each country. In this step, principle component analysis was performed to reveal the optimal number of factors, followed by exploratory factor analysis to extract the latent factors (Table 3a, b). Secondly, the structure of each component was verified through confirmatory factor analysis. Lastly, the verification of the model was conducted through SEM.

In both cases, only one factor was generated from the affective image scale (communalities from 0.34, and factor loadings near 0.60 and above). The Cronbach's alphas of this scale are 0.761 and 0.750 for Malaysia and Thailand respectively, with all corrected total-item correlation values > 0.30 (Leech, Barret & Morgan 2005). Three factors were extracted from the fourteen cognitive items for both Malaysia and Thailand; however, the structure of this component differs between the two countries.

Table 3a. Principle axis factoring solution for the images of Malaysia

	Factor	Items	Loading	Communality	Eigenvalue	% of variance	Crobach's alpha
Cognitive scale [1]	Cognitive Factor 1				2.145	23.831	.792
		C5	.577	.381			
		C9	.579	.403			
		C10	.564	.333			
		C12	.678	.512			
		C13	.729	.590			
	Cognitive Factor 2				1.190	13.225	.653
		C3	.478	.301			
		C8	.882	.824			
	Cognitive Factor 3				1.159	12.883	.715
		C6	.772	.658			
		C7	.685	.493			
Affective scale [2]	Affective				1.740	43.510	.761
		A1	.715	.511			
		A2	.586	.343			
		A3	.647	.418			
		A4	.684	.468			

Note. [1] Determinant = 0.085; Kaiser-Meyer-Olkin = 0.782; Chi-square = 752.978; Significance = 0.000
[2] Determinant = 0.392; Kaiser-Meyer-Olkin = 0.738; Chi-square = 288.053; Significance = 0.000

In the case of Malaysia, C1, C2, C4, C11, and C14 were removed due to high crossloading: the difference between two loadings > 0.20 with the cutting point set at 0.40 (Ferguson & Cox 1993). All the three factors have eigenvalues > 1, and they account for 49.940% of the total variance; C3 has the lowest communality value (0.301), and the lowest factor loading value (0.478). The internal consistency levels of the three factors are 0.792 (factor 1: C5, C9, C10, C12, and C13), 0.653 (factor 2: C3, and C8), and 0.715 (factor 3: C6, and C7) (corrected total-item correlations > 0.30). Although the second factor has an alpha < 0.70, it is considered to be sufficient in the exploratory stage (Nunnally & Bernstein 1994; Chi & Qu 2008). To verify the latent construct, confirmatory factor analysis was conducted. All the items significantly load on their factors (factor loadings range from 0.552 to 0.902); thus, the convergent validity is achieved (Kline 1998). Based on the data provided by the analysis program, average variance extracted (AVE), and construct reliability (CR) were calculated for each factor to measure the validity (Fornell & Larcker 1981). The AVEs (CRs) of the three factors are: 0.498 (0.831) for factor 1, 0.666 (0.794) for factor 2, and 0.491 (0.658) for factor 3. AVE should not be less than 0.50 (Fornell & Larcker 1981), and CR should exceed 0.70 (as a rule of thumb). Thus, the CRs of the three factors meet the criterion, while AVEs are close to the threshold of 0.50. To examine the discriminant validity of the cognitive image construct, the squared variance (SV) of the correlations between two factors was compared to the two factors' AVEs. As the results, all the SVs have smaller values than the AVEs; thus, this criterion of discriminant validity is met (Fornell & Larcker 1981). The model also has acceptable fit indices: χ 2 /df = 1.967, SRMR = 0.034, RMSEA = 0.056, NFI = 0.938, CFI = 0.968, GFI = 0.968, and AGFI = 0.941. Consequently, the structure of the cognitive image of Malaysia is confirmed.

With Thailand, the three cognitive factors make up 41.879% of the total variance, and all of them have eigenvalues > 1. C8 and C9 have communality values < 0.3 (0.271 and 0.286). However, their respective factor loading values are 0.428 (C8) and 0.460 (C9); thus, they were kept for confirmatory factor analysis. In this case, the items with high cross-loading were removed include C1, C10, and C12. The Cronbach's alphas of the three factors are 0.693 (factor 1: C6, C7, C11, and C14), 0.679 (factor 2: C4, C5, C9, and C13), and 0.703 (factor 3: C2, C3, and C8) (corrected total-item correlations > 0.30). Confirmatory factor analysis was conducted later to confirm the structure of the cognitive image construct. Model fit indices of the initial model includeχ 2 /df = 3.132, SRMR = 0.065, RMSEA = 0.083, NFI = 0.840, CFI = 0.883, GFI = 0.932, and AGFI = 0.890. Following the suggestion made by the analysis program, a path which correlates the unobserved exogenous variables of two items C11 and C14 was added. Consequently, the fits were improved: χ 2 /df = 2.752, SRMR = 0.066, RMSEA = 0.075, NFI = 0.863, CFI = 0.906, GFI = 0.942, and AGFI = 0.904. All the items have significant loadings on their factors (0.412 – 0.789). In addition, two indices were calculated to measure the validity of the construct: average variance extracted (AVE), and construct reliability (CR) (Fornell & Larcker 1981). AVEs (CRs) of the three factors are 0.380 (0.700), 0.431 (0.752), and 0.500 (0.745) respectively. The first two factors have low AVEs (compared to the 0.50 criterion), although they have acceptable level of CRs and the squared variances don't exceed the AVE values.

Table 3b. Principle axis factoring solution for the images of Thailand

	Factor	Items	Loading	Communality	Eigenvalue	% of variance	Crobach's alpha
Cognitive scale [1]	Cognitive Factor 1				1.556	14.145	.693
		C6	.553	.372			
		C7	.609	.388			
		C11	.602	.380			
		C14	.546	.407			
	Cognitive Factor 2				1.543	14.028	.679
		C4	.505	.338			
		C5	.677	.470			
		C9	.460	.286			
		C13	.586	.420			
	Cognitive Factor 3				1.508	13.705	.703
		C2	.562	.444			
		C3	.889	.829			
		C8	.428	.271			
Affective scale [2]	Affective				1.810	45.258	.750
		A1	.755	.570			
		A2	.624	.389			
		A3	.606	.367			
		A4	.696	.484			

Note. [1] Determinant = 0.075; Kaiser-Meyer-Olkin = 0.788; Chi-square = 792.394; Significance = 0.000
[2] Determinant = 0.365; Kaiser-Meyer-Olkin = 0.744; Chi-square = 310.169; Significance = 0.000

The confirmation of the theoretical model for each country was followed in the last step. The fit indices of the initial model in the case of Thailand are $\chi 2$ /df = 6.476, SRMR = 0.174, RMSEA = 0.133, NFI = 0.541, CFI = 0.577, GFI = 0.801, and AGFI = 0.738. These when testing the model for Malaysia are $\chi 2$ /df = 6.232, RMR = 0.236, RMSEA = 0.168, NFI = 0.623, CFI = 0.660, GFI = 0.833, and AGFI = 0.775. The modifications made later could not improve the model fits for both countries. Consequently, in this analysis solution, the theoretical model is not approved (Figure 3a for Malaysia, and Figure 3b for Thailand).

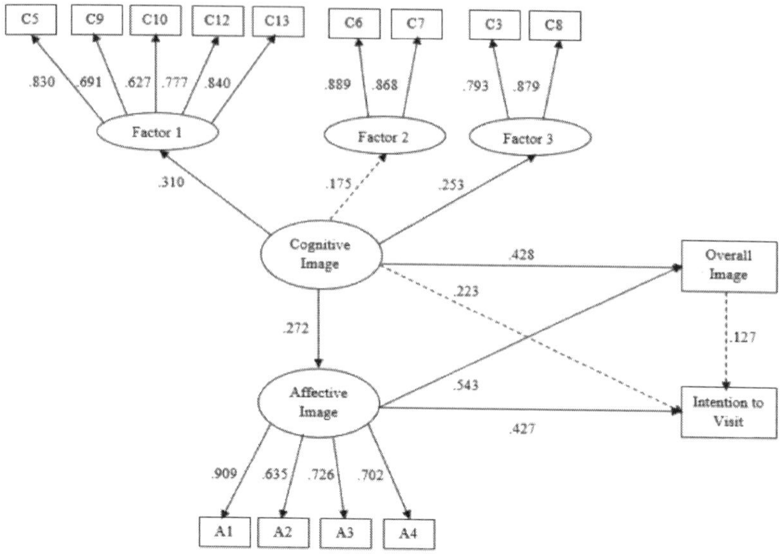

(3a)

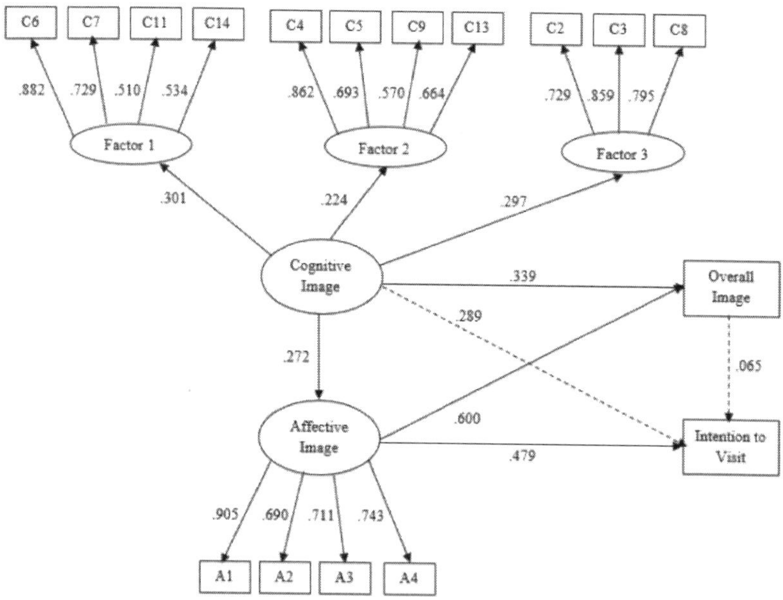

(3b)

Figure 3. The structural model tested for (3a) Malaysia, and (3b) Thailand

4.2.2 Second analysis solution

In the second solution, the whole scale of the cognitive image was used in the analysis. As the results, the fit indices when testing the theoretical model for Thailand are $\chi 2$ /df = 3.530, SRMR = 0.075, RMSEA = 0.090, NFI = 0.715, CFI = 0.775, GFI = 0.822, and AGFI = 0.774. Those in the Malaysia's case include $\chi 2$ /df = 3.430, SRMR = 0.070, RMSEA = 0.089, NFI = 0.741, CFI = 0.799, GFI = 0.831, and AGFI = 0.786. These values are close to the acceptable thresholds of fit indices.

Thus, following the modification indices suggested by AMOS, correlations were added between eleven pairs of unobserved exogenous variables in both cases. The new fit indices are $\chi 2$ /df = 1.987, SRMR = 0.057, RMSEA = 0.056, NFI = 0.850, CFI = 0.918, GFI = 0.905, and AGFI = 0.871 for Thailand; and $\chi 2$ /df = 1.880, SRMR = 0.053, RMSEA = 0.053, NFI = 0.867, CFI = 0.932, GFI = 0.911, and AGFI = 0.880 for Malaysia. Consequently, after adjusting, both models have acceptable fit indices (except NFI and CFI), and Malaysia's model has slightly better fits as compared to Thailand's model.

The loadings of the items, and the effects of the hypothesized relationships are provided in Table 4. The cognitive scale has an alpha of 0.854 for Malaysia, and 0.829 for Thailand. All the items significantly load on the cognitive image construct; however, the loadings are quite low (0.358 – 0.686 for Malaysia, and 0.397 – 0.679 for Thailand). The loadings of the affective items on the affective image construct are higher (0.516 – 0.797 for Malaysia, and 0.584 – 0.803 for Thailand), with the alpha of the scale is 0.750 for Malaysia, and 0.761 for Thailand. The path assuming the effect that cognitive image has on affective image has a standardized regression weight (β) around 0.75, significant level (p) < 0.000 for both countries. The effect that affective image has on overall image is represented by a $\beta = 0.586$ in Malaysia's case, and 0.554 in Thailand's case ($p < 0.000$). Affective image also has significant influence on intention-to-visit: $\beta = 0.367$, $p < 0.01$ for Malaysia, and $\beta = 0.518$, $p < 0.000$ for Thailand

These findings only support some of the hypotheses. It was revealed that cognitive image significantly affected affective image, affective image significantly influenced overall image, and both affective and overall images affected intention-to-visit at significant levels. Consequently, H1a/c, and H2b/c were supported. This is consistent with previous studies' findings. However, the effect that cognitive image has on overall image and intention-to-visit could not be confirmed in the contexts of both Malaysia and Thailand due to small effect sizes and insignificance levels (p > 0.05). Thus, H1b, and H2a were rejected. The similar finding was found in Lin, Morais, Kerstetter and Hou (2007), in which in the case of Taiwanese potential travelers' perceptions of the theme parks, cognitive image didn't influence overall image. Moreover, in the context of Japanese Halyu fans' perceptions of and intention to visit Korea, affective image was revealed to have a larger effect on intention-tovisit as compared to cognitive image (Lee, Scott & Kim 2008). In addition, this study has confirmed the importance of emotional perceptions to the formation of destination image as seen in previous studies

Table 4. Results of the second analysis solution

Observed, endogenous variables	Loading on		Crobach's alpha		Effect on affective image		Effect on overall image		Effect on intention to visit	
	M	T	M	T	M	T	M	T	M	T
	Cognitive		.854	.829	.757	.745	.061[1a]	.138[1b]	.003[2a]	-.071[2b]
C1 (e1)*	.592	.679								
C2 (e2)	.599	.529								
C3 (e3)	.490	.435								
C4 (e4)	.492	.397								
C5 (e5)	.575	406								
C6 (e6)	.425	.428								
C7 (e7)	.358	.418								
C8 (e8)	.512	.419								
C9 (e9)	.566	.499								
C10 (e10)	.514	.480								
C11 (e11)	.422	.498								
C12 (e12)	.649	.559								
C13 (e13)	.686	.620								
C14 (e14)	.558	.645								
	Affective		.750	.761			.586	.554	.367[3a]	.518
A1 (e15)	.797	.803								
A2 (e16)	.516	.584								
A3 (e17)	.647	.621								
A4 (e18)	.603	.636								
O (e19)									.249	.173[3b]
I (e20)										

Note. M = Malaysia; T = Thailand

* Unobserved, exogenous variables were put in the parentheses.

1a. $p = 0.563$; 1b. $p = 0.138$; 2a. $p = 0.980$; 2b. $p = 0.461$; 3a. $p = 0.003$; 3b. $p = 0.014$. Other estimates are significant at the 0.000 level.

In the case of Malaysia, the following associations were incorporated into the adjusted model: e1-e2, e2-e3, e3-e8, e4-e5, e5-e11, e6-e7, e7-e17, e9-e10, e11-e14, e12-e13, and e16-e18. In the case of Malaysia, the following associations were incorporated into the adjusted model: e1-e13, e2-e3, e3-e8, e4-e5, e5-e11, e6-e7, e7-e16, e7-e20, e9-e11, e14-e16, and e16-e18.

The modifications and their results show that there are several associations between the variables of the theoretical model are hidden.

This is reasonable since the whole scale of each construct was used in SEM. The large number of variables in the cognitive image construct (14) has resulted in a number of unobserved correlations.

The testing of the structural model in both cases and both solutions seems supporting the findings of the comparison of two countries' images. The outcomes of the tested models are almost similar in terms of the sizes and direction of the effects. Thus, with the Japanese potential tourists subjected in this study, the images of Malaysia and Thailand were not much differentiated.

5. CONCLUDING REMARKS

The purpose of this paper was to investigate the images of two tourism destinations in the Southeast Asia region (i.e., Malaysia, and Thailand) as perceived by potential youth tourists in the Japanese market. The findings revealed that Japanese undergraduate students perceived the images of Malaysia more favorable as compared to Thailand. There were more similarities between the images of the two countries compared to the differences, including their affective images. However, the role of the emotional/affective perceptions should not be overlooked since they are important for the formation of both perceived destination images and intention-to-visit.

The affective perceptions reflect one's feelings toward the destination. The formation of affective images is affected by the cognitive images, or the attributes of the destination. In case of potential tourists (especially those who haven't visited the destination), these images are formed through the encounter with information sources. Thus, the projection and promotion of a destination's attributes not only shape the cognitive images held by the potential tourists, but also influence the formation of affective images. Affective images, in its turn, influence potential tourists' intention-to-visit. The whole process seems to be complicated and intrinsic. What the marketers/managers of a destination can do is to create and manage positive information about their destination, especially during the recession and/or crisis periods

In this study, Thai affective images' effect on the respondents' intention-to-visit the country ($\beta = 0.518$, $p < 0.000$) was larger than that of Malaysia ($\beta = 0.367$, $p < 0.01$). Japanese undergraduate students also showed more intention to visit Thailand than to Malaysia. Thus, although Thai images were not evaluated as highly as were Malaysia's images, Thailand was more likely to be considered as a potential destination by the Japanese respondents. Thai tourism may want to use this "emotional advantage" to attract Japanese student/youth tourists to their destination. In this case, the efforts to distinguish Thai images from those of Malaysia may not be necessary. However, it is significant to maintain and/or improve the current status of the cognitive images to strengthen the emotional ties, and consequently to increase the intention-to-visit of these potential tourists.

It should be noted that the images held by the majority of the respondents of this study were formed before they actually visit the destinations, or were the secondary image (Lopes 2011). Thus, the information sources and their quality are important in forming the pre-visit images of the Japanese student respondents. In recent studies, the Internet was reported to be the most important source of information for youth travelers, followed by travel magazines (e.g., Hanawa 2009; Okamoto 2009; Kim & Kamata 2010). Moreover, Japanese youth prefer to access the Internet through their mobile phones rather than through private computers (Schiano, Elliott & Bellotti 2006). This characteristic should be considered in order to design the most suitable strategies to communicate with the target audience.

One limit of this study is its probabilistic generalizability (Blair & Zinkhan 2006). Since only the undergraduate students were approached, the results cannot be generalized to the whole Japanese markets. However, this study introduces several recommendations for future research. Other destinations in the region (e.g., Indonesia, the Philippines, Vietnam), other generating markets (e.g., China, Europe, America), and other sub-markets in Japan (e.g., seniors) may be targeted to replicate and generalize the theoretical model and findings of this study (Blair & Zinkhan 2006). Another possibility is to measure the images of the Southeast Asia region as a single destination. Southeast Asia is on their way to build the regional community. In this context, the understanding of the images of the whole region will support both the collective (regional) and individual (national) efforts to develop its tourism industry, and to reach the goal of the establishment of the economic community in 2015. In addition, future studies may continue the current effort to see how the recent incidents which has happened in the two countries (the aviation incidents in Malaysia, and the political instability in Thailand) influence the perceptions of their potential tourists. The destination marketers in

Malaysia and Thailand will be provided some valuable inputs for the management of their respective country's images during and after crises.

REFERENCES

1. Andersen, V., Prentice, R. & Watanabe, K. 2000, Journey for experiences: Japanese independent travelers in Scotland, *Journal of Travel and Tourism Marketing* 9 (1/2), 129-151. Baloglu, S. 1996, *An empirical investigation of determinants of tourist destination image*, Doctoral Dissertation, Virginia Polytechnic Institute and State University, America.

2. Baloglu, S. & McCleary, K.W. 1999a, A model of destination image formation, *Annals of Tourism Research* 26 (4), 868-897.

3. Baloglu, S. & McCleary, K.W. 1999b, U.S. international pleasure travelers' images of four Mediterranean destinations: A comparison of visitors and non-visitors, *Journal of Travel Research* 38, 144-152.

4. Blair, E. & Zinkhan, G.M. 2006, Nonresponse and generalizability in academic research, *Journal of the Academy of Marketing Science* 34 (1), 4-7.

5. Bozbay, Z. & Ozen, H. 2008, Measuring the destination images of European cities, *Proceedings of the 4th world conference for graduate research in tourism, hospitality and leisure*, Anatolia, Antalya, pp. 725-738.

6. Bui, L.H. 2011, Congruency between the projected and perceived tourism destination image of Vietnam, *Journal of International Business Research* 10 (2), 1-13.

7. Chang, T.C. 2004, *Tourism in a "borderless" world: The Singapore experience*, East-West Center, Honolulu.

8. Chen, J.S. 1996, *Factors influencing tourists' choice of heritage destinations*, Doctoral Dissertation, The Graduate School, School of

Hotel, Restaurant and Recreation Management, The Pennsylvania State University, America.

9. Chena, H.-J., Chen, P.-J. & Okumus, F. 2013, The relationship between travel constraints and destination image: A case study of Brunei, *Tourism Management* 35, 198-208.

10. Chens, C.-Y., Sok, P. & Sok, K. 2008, Evaluating the competitiveness of the tourism industry in Cambodia: Self-assessment from professionals, *Asia Pacific Journal of Tourism Research* 13 (1), 41-66.

11. Chi, C.G.-Q. & Qu, H. 2008, Examining the structural relationships of destination image, tourist satisfaction and destination loyalty: An integrated approach, *Tourism Management* 29, 624-636.

12. Chiang, Y.-S., Lu, J.-L. & Chang, H.-C. 2003, Modeling the effect of destination attributes on the intercity travelers' mode choice behavior in Taiwan area, *Proceedings of the Eastern Asia Society for Transportation Studies* 4, 717-730.

13. Crompton, J. 1992, Structure of vacation destination choice sets, *Annals of Tourism Research* 19, 420-434.

14. Crompton, J.L. 1979, Motivations for pleasure travel, *Annals of Tourism Research* 6, 408-424. de Guzman, A.B., de Castro, B.V., Calanog, J.F., Taguinin, A.J., Afalla, J.R., Aldover, A.L. & Gotangco, M.T. 2012, The Australian tourists' travel motivation and pre- and post-images of The Philippines as their destination, *Asia-Pacific Journal of Innovation in Hospitality and Tourism* 1 (2), 143-164.

15. Echtner, C.M. & Ritchie, J.R. 2003, The meaning and measurement of destination image. *The Journal of Tourism Studies* 14 (1), 37-48.

16. Ferguson, E. & Cox, T. 1993, Exploratory factor analysis: A users' guide, *International Journal of Selection and Assessment* 1 (2), 84-94.

17. Fornell, C. & Larcker, D.F. 1981, Evaluating structural equation models with unobservable variables and measurement error, *Journal of Marketing Research* 18 (1), 39-50.

18. Gartner, W. 1993, Image formation process, *Journal of Travel and Tourism Marketing* 2 (2/3), 191-215.

19. Hanawa, I. 2009, A study on intangibility in tourism industry: New communication channel, *Journal of Commerce* 30, 67-84. (In Japanese)

20. Henderson, J.C. 2007, Uniquely Singapore? A case study in destination branding, *Journal of Vacation Marketing* 13 (3), 261-274.

21. Howard, R.W. 2009, Risky business? Asking tourists what hazards they actually encountered in Thailand, *Tourism Management* 30, 359-365.

22. Hyun, M.Y. & O'Keefe, R.M. 2012, Virtual destination image: Testing a telepresence model, *Journal of Business Research* 65, 29-35.

23. ITB Berlin. 2013, *ITB world travel trends report*, Messe Berlin GmbH, Berlin.

24. Japan Tourism Board [JTB] Corp. 2010, *JTB report 2010 - All about Japanese overseas travel*, JTB Corp, Tokyo.

25. Kamil, A.A. 2010, Travelers' perceptions of Malaysia as their next holiday destination, *Ozean Journal of Social Sciences* 3 (2), 171-183.

26. Kim, H. & Kamata, H. 2010, Youth's awareness of travel, *Economics Research* 188, 177-191. (In Japanese)

27. Kim, S. & Yoon, Y. 2003, The hierarchical effects of affective and cognitive components on tourism destination image, *Journal of Travel & Tourism Marketing* 14 (2), 1-22.

28. Kline, R.B. 1998 *Principles and practices of structural equation modeling*, The Guilford Press, New York.

29. Kwon, H.-J. 2002, *The effects of prototypicality on vacation destination choice behavior*, Doctoral Dissertation, The Graduate School of Clemson University, America.

30. Le, T.A. 2008, Tourism destination image: Vietnam as a tourism destination for Japanese travelers, *Proceedings of the Forth Vietnam Development Forum – Tokyo on The Development of Vietnam*, National Graduate Institutes for Policy Studies, Tokyo, pp. 180-196. Lee, C. 2002, Analysis of images of Huis Ten Bosch and their effects on tourism: The case of college students in Korea, *Nagasaki International University Series* 2, 19-25. (In Japanese) Lee, S., Scott, D. & Kim, H. 2008, Celebrity fan involvement and destination perceptions, *Annals of Tourism Research* 35 (3), 809-832.

31. Leech, N.L., Barret, K.C. & Morgan, G.A. 2005, *SPSS for intermediate statistics: Use and interpretation* (2 ed.). Lawrence Erlbaum Associates, Mahwah, NJ.

32. Lin, C.-H., Morais, D.B., Kerstetter, D.L. & Hou, J.-S. 2007, Examining the role of cognitive and affective image in predicting choice across natural, developed, and theme park destinations, *Journal of Travel Research* 46, 183-194.

33. Litvin, S.W. & Ling, S.N. 2001, The destination attribute management model: An empirical application to Bintan, Indonesia, *Tourism Management* 22, 481-492.

34. Lopes, S.D. 2011, Destination image: Origins, developments and implications, *PASOS- Revista de Turismo y Patrimonio Cultural [Journal of Tourism and Cultural Heritage]* 9 (2), 305-315.

35. Lumsdon, L. 1997, *Tourism marketing*, International Thomson Business Press, London. Miyamori, M. 2001, A study of the public relations and the advertising activities of the Okinawa summit for developing the sight-seeing image of Okinawa globally, *Journal of Business and Economics* 30 (1), 67-82. (In Japanese)

36. Mohamad, M., Abdullah, A.R. & Mokhlis, S. 2012, Tourists' evaluations of destination image and future behavioral intention: The case of Malaysia, *Journal of Management and Sustainability* 2 (1), 181-189.

37. Molina, M.A., Frias-Jamilena, D.-M. & Castaneda-Garcia, J.A. 2013, The moderating role of past experience in the formation of a tourist destination's image and in tourists' behavioral intentions. *Current Issues in Tourism* 16 (2), 107-127.

38. Nicoletta, R. & Servidio, R. 2012, Tourists' opinions and their selection of tourism destination images: An affective and motivational evaluation. *Tourism Management Perspectives* 4, 19-27.

39. Nunnally, J.C. & Bernstein, I.H. 1994, *Psychometric theory* (3 ed.), McGraw-Hill, New York. Okamoto, T. 2009, Tourism in the information revolution era: Travel informationalization generation, *Web-Journal of Tourism and Cultural Studies* 6, 1-16. (In Japanese)

40. Page, S. 2003, *Tourism management – managing for change*, Butterworth-Heinemann, Amsterdam.

41. Pawitra, T.A, & Tan, K.C. 2003, Tourist satisfaction in Singapore – A perspective from Indonesian tourists, *Managing Service Quality* 13 (5), 399-411.

42. Reisinger, Y. & Turner, L. 2000, Japanese tourism satisfaction: Gold Coast versus Hawaii, *Journal of Vacation Marketing* 6 (4), 299-317.

43. Richter, L. 2009, Tourism policy-making in Southeast Asia: A twenty-first century perspective. In M. Hitchcok, V.T. King & M. Parnwell (eds.), *Tourism in Southeast Asia – Challenges and new directions*, University of Hawaii Press, Honolul, pp. 132-145.

44. Rittichainuwat, B.N. & Chakraborty, G. 2009, Perceived travel risks regarding terrorism and disease: The case of Thailand, *Tourism Management* 30, 410-418.

45. Rittichainuwat, B.N., Qu, H. & Mongknonvanit, C. 2002, A study of the impact of travel satisfaction on the likelihood of travelers to revisit Thailand, *Journal of Travel & Tourism Marketing* 12 (2/3), 19-43.

46. Rittichainuwat, B.N., Qu, H. & Mongkhonvanit, C. 2008, Understanding the motivation of travelers on repeat visits to Thailand, *Journal of Vacation Marketing* 14 (1), 5-21.

47. Russell, J.A. & Pratt, G. 1980, A description of the affective quality attributed to environments, *Journal of Personality and Social Psychology* 38 (2), 311-322.

48. San Martin, H. & del Bosque, I.A. 2008, Exploring the cognitive-affective nature of destination image and the role of psychological factors in its formation, *Tourism Management* 29, 263-277.

49. Sangpikul, A. 2008, Travel motivations of Japanese senior travelers to Thailand, *International Journal of Tourism Research* 10, 81-94.

50. Schermelleh-Engel, K., Moosbrugger, H. & Müller, H. 2003, Evaluating the fit of structural equation models: Tests of significance and descriptive goodness-of-fit measures, *Methods of Psychological Research Online* 8 (2), 23-74.

51. Schiano, D.J., Elliott, A. & Bellotti, V. 2006, Tokyo youth at leisure: Towards the design of new media to support leisure planning and practice, *Computer Supported Cooperative Work* 16 (1/2), 45-73.

52. Seddighi, H.R., Nuttall, M.W. & Theocharous, A.L. 2001, Does cultural background of tourists influence the destination choice? An empirical study with special reference to political instability, *Tourism Management* 22, 181-191.

53. Stepchenkova, S. & Mills, J.E. 2010, Destination image: A meta-analysis of 2000-2007 research, *Journal of Hospitality Marketing & Management* 19, 575-609.

54. Talib, H. 2005, *A cross-cultural comparison of local and foreign tourist preferences towards selected visual images of Malaysian landscapes*, Unpublished Master's Thesis, The School of Graduate Studies, Universiti Putra Malaysia, Selangor Darul Ehsan, Malaysia.

55. Tan, A.Y., McCahon, C. & Miller, J. 2002, Modeling tourist flows to Indonesia and Malaysia, *Journal of Travel & Tourism Marketing* 13 (1/2), 61-82.

56. Truong, T.H. & King, B. 2006, Comparing cross-cultural dimensions of the experiences of international tourists in Vietnam, *Journal of Business Systems, Governance and Ethics* 1 (1), 65-75.

57. Truong, T.-H. 2005, Assessing holiday satisfaction of Australian travelers in Vietnam: An application of the HOLSAT model, *Asia Pacific Journal of Tourism Research* 10 (3), 227-246. Wang, K.-L. & Wu, C.-S. 2003, A study of competitiveness of international tourism in the Southeast Asian region, *Trade in Services in the Asia Pacific Region* 11, 315-345.

58. Weaver, D. & Lawton, L. 2002, *Tourism management* (2 ed.), John Wiley & Sons Australia, Sydney.

59. World Tourism Organization [UNWTO] 2007, *A practical guide to tourism destination management*, UNWTO, Madrid.

60. World Tourism Organization [UNWTO] 2008, *Youth travel matters – Understanding the global phenomenon of youth travel*, UNWTO, Madrid.

61. Yamashita, S. 2009, Southeast Asian tourism from a Japanese perspective. In M. Hitchcok, V.T. King & M. Parnwell (eds.) *Tourism in Southeast Asia – Challenges and new directions*, University of Hawaii Press, Honolulu, pp. 189-205.

62. Yeung, S., Wong, J. & Ko, E. 2004, Preferred shopping destination: Hong Kong versus Singapore, *International Journal of Tourism Research* 6, 85-96.

10

AN EXAMINATION OF ANTECEDENTS OF LOYALTY INTENTION TO TRAVEL: THE CASE OF MALAYSIA

Abbas Alizadeh, Golnaz Saghafi

University Kebangsaan Malaysia, Graduate School of Business, Malaysia.

ABSTRACT

The tourism industry has experienced a rapid and continuous growth during the last decade, making it one of the leading industries of the world. As such, destination marketers increasingly seek to understand the various constructs that determine tourists' destination loyalty. This study aims at examining the antecedents of travelers' loyalty toward a destination through an integrated tourist behavior model. The model used for this research was developed based on the existing literature on travelers' behavioral intentions. The study tested thirteen hypotheses on factors which impact the travelers' loyalty toward a destination with the empirical data through structured questionnaires from a sample of 202 tourists in Malaysia, the city of Kuala Lumpur. The results from multiple regression analysis revealed that (1) food image, destination image and information sources influenced trip quality; (2) the destination image is an antecedent of perceived trip value; (3) both trip quality and perceived trip value positively impact on tourists' satisfaction and behavioral intention; (4) satisfaction is a predictor of behavioral intention; and (5) tourist's behavioral intention to visit the destination positively affects tourist's destination loyalty. The study contributes to the current body of literature by offering an integrated model of tourist loyalty. Based on the findings of this study, theoretical insights were offered for researchers and practical implications were proposed to encourage destination marketers to direct their resources on the right efforts that can lead to tourists' loyalty.

KEYWORDS

Destination image, Destination loyalty, Food image, Tourist satisfaction, Visit intention

1. INTRODUCTION

The tourism industry is one of the largest businesses worldwide (Sipe, 2013) and plays a significant role in the economic sector (Baker, 2014; Liu, Tzeng, & Lee, 2012). Due to its growth, competition among tourism destinations has been increasing. As a result, companies try to promote their destinations by building a positive image for their customers - called desired perception.

You (2011) noted that one primary objective of tourism companies is to achieve customer loyalty. The stiff competition in this industry has led to improvements in the products and services related to tourists (Lin, 2006). Furthermore, these developments can improve the destinations' economical level, image and government revenue. It is therefore essential to attract and retain tourists to revisit and make them loyal to the destination (Chen & Tsai, 2007).

Although many researchers (e.g. Chi & Qu, 2008; Yang & Peterson, 2004; You, 2011) have proposed different models of customer loyalty in tourism context, the lack of an integrated approach is evident. For instance, the focus of several studies is on examining the behavioral intention without considering the actual intention to revisit a destination. Other studies focus on the role of travelers' satisfaction and neglect the tourists' intention to visit and revisit a destination. Finally, some studies are focused on the tourists' loyalty toward a destination while missing some important antecedents of loyalty such as information sources and perceived value (Cronin Jr et al., 2000; Žabkar et al., 2010; Ranasinghe, 2012). To address this gap, this study examines the role of critical elements of destination loyalty including destination image, food image, information sources, trip quality, perceived value, satisfaction, and behavioral intentions. The study tests thirteen hypotheses on these factors which ultimately impact the travelers' loyalty toward a destination with the empirical data through structured questionnaires from a sample of 202 tourists in Malaysia destination. We believe that by understanding the relationships between these elements which are critical for the development of the tourism in a given destination, managers would make better and more effective decisions and they would be able to use the resources more efficiently.

2. RESEARCH FRAMEWORK AND HYPOTHESES

Several models have been developed to understand how travelers remain loyal to a destination (e.g. Rajesh, 2013; Chi & Qu, 2008; You, 2011; McDougall & Levesque, 2000; Yang & Peterson, 2004; Badarneh & Som, 2011; Feng, 2007; Murphy et al., 2000; Park & Nunkoo, 2013). The purpose of the current model is to understand and examine the factors which impact tourists' loyalty toward Malaysia as a destination. The current study aims to focus on travelers intention to revisit a destination by investigating and examining the impact of its antecedents including food image, destination image, information sources, trip quality, perceived value, satisfaction and behavioral intention. The present research investigates the travelers who have visited Malaysia.

Based on the literature, an integrated model has been created for this research paper (see Figure 1). This theoretical framework combines the model of tourists' behavioral intentions by Chen and Tsai (2007), model of revisit intention by Badarneh and Som (2011) and culinary tourism model by Ab Karim (2006).

2.1 Food image

Food is a basic need for human to survive, but this basic need is very critical in the tourism industry. In this industry, creating good service and positive food image is very important. Food image is a mental perception and feeling of a person toward food. It is important to mention that the importance of food and food industry has increased over the past few years. Food is used as an eye-catcher element in the tourism industry and different programs such as videos, television, etc. In fact, this element can attract many tourist everywhere (Hjalager &

Corigliano, 2000). Eating out is an activity which can bring pleasure and enjoyment for people who are in the travel and when they are away from home and is considered as a crucial part of the journey (Ab Karim, 2006).

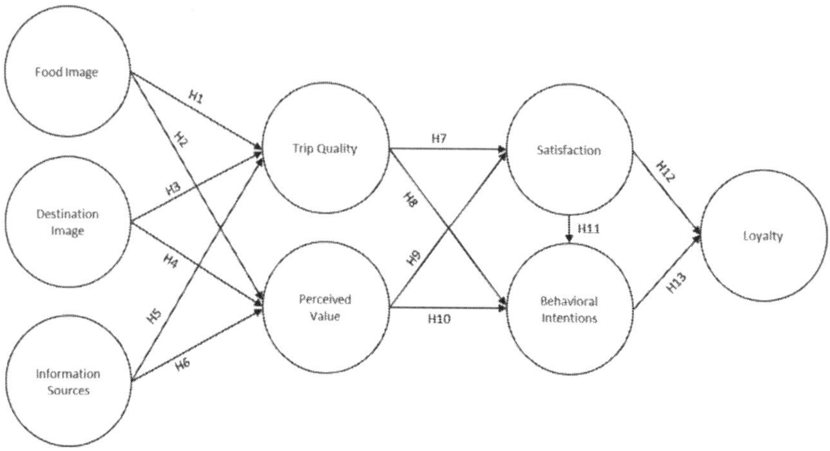

Figure 1. The proposed conceptual framework

As Chen and Tsai (2007) argue, both food and beverage affect the hospitality and amenity which considered as important factors that build trip quality (Chen & Tsai, 2007). However, tourism organizations have to offer different and various kinds of food of high quality to be able to compete in this industry. Offering various local foods is one way to introduce a tourist destination (Ab Karim, 2006).

Hypothesis 1. Food image of the destination positively affects tourist's trip quality.

In a tourism context, food image is one of the essential characteristics of service quality (Good et al., 2008) and many researchers argue that service quality positively influences perceived value (e.g. Kuo et al., 2009; Bauer et al., 2006). Therefore, it is suggested that food image can influence a tourist's perceived value.

2.2 Hypothesis 2. Food image of the destination positively affects tourist's perceived trip value. Destination image

As Chen and Tsai (2007) state, destination image is defined as individual's knowledge and feelings about a specific destination (Chen & Tsai, 2007). There is an agreement on the importance of destination image in the tourism industry, but a few researches have been conducted on this critical factor (Badarneh & Som, 2011). This important factor itself includes exogenous factors such as natural environment and endogenous factors such as tourist's satisfaction (Chi & Qu, 2008).

Chen and Tsai (2007) have conducted a study related to destination image and trip quality. The results show that destination image has a direct impact on the trip quality. As such, the more favorable the destination image, the higher the trip quality.

Hypothesis 3. Destination image positively affects tourist's trip quality.

Moreover, the destination image affect the tourist's perceived value of a destination. It means that when a destination image is favorable, it results in the higher perceived value. The perceived value can be measured by three items which are money value, time value and effort value (Chen & Tsai, 2007). To conclude, the more favorable the destination is, the more a tourist values his/her effort, time and money spent for visiting the destination.

Hypothesis 4. Destination image positively affects tourist's perceived trip value.

2.3 Information sources

Nowadays, travelers seek for information which helps them to make a better decision for choosing a destination to travel. The type of the information they look for, depends on the credibility of the information source (Ab Karim, 2006). The information that tourists receive come from various sources and they are divided into two groups. First group is informal information sources such as word of mouth or the information that people get from their family and friends. Second group is formal information sources such as commercials or electronic media.

It is important to mention that the type of traveler (e.g. cultural tourist, sport tourist etc.) and travel objectives determine the type of required information (Ab Karim, 2006).

It is argued that travelers who try to collect more information before choosing a destination to travel, will experience a better trip quality (McIntosh et al., 1995). Travelers can be classified based on the level of information search and the information type they receive before travel. Travelers who get a high level of information are more satisfied with their trip quality (Snepenger et al., 1990). It is worthwhile to mention that demographics characteristics including gender, age, ethnicity, knowledge of languages, employment status, and even location have an important effect on travelers' decision making and choosing information source (Ab Karim, 2006).

Hypothesis 5. The type of information source has a significant effect toward the tourist's trip quality.

In addition, in a research by Gruen et al. (2006), it was found that electronic word-of-mouth, which is a type of information source, not only influences perceived value but also can influence loyalty intentions. Hence, it is suggested that the type of information source affects perceived value of customers.

Hypothesis 6. The type of information source has a significant effect toward the tourist's perceived trip value.

2.4 Trip quality

The term quality becomes meaningful when the performance of a quality or service is same as or more than the customer's expectation. In the tourism industry, a trip quality is existed when the traveler has positive feelings towards his/her trip (Murphy et al., 2000).

Žabkar et al. (2010) offer a model that shows the quality of the trip has a positive effect on both travelers' satisfaction and behavioral intention. The results confirmed that the trip quality has a direct impact on the tourists' satisfaction and behavioral intention.

Hypothesis 7. Trip quality positively affects tourist's travel satisfaction.

Moreover, it is argued that satisfaction mediates the relationship between service quality and behavioral intentions (Cole & Illum, 2006; Chi & Qu, 2008). However, the results of Žabkar et al.'s (2010) research confirms that perceived quality of destination's offerings can directly and positively influences tourist's behavioral intention.

Hypothesis 8. Trip quality positively affects tourist's behavioral intention to visit the destination.

2.5 Perceived value

Oliver and DeSarbo (1988) discuss that perceived value can be considered as the ratio of customer's output/input to producer's output/input (Oliver & DeSarbo, 1988). In fact perceived value is a customer's evaluation about the characteristics of a product or service (Bolton & Lemon, 1999). In a tourism context, the traveler evaluates both monetary and nonmonetary factors such as the time, energy and effort that have been spent for the trip (Yang & Peterson, 2004). In addition, Sweeney and Soutar (2001) have proposed a model and argue that a customer evaluates and judges according to the perception of what he/she has got and given (Sweeney & Soutar, 2001).

McDougall and Levesque (2000) discuss that both perceived value and service quality are variables which have to be considered as drivers of satisfaction. In their study it has been demonstrated that customer perception of the service quality and his/her evaluation of the service directly affect the feeling of satisfaction. Moreover the results confirmed that perceived value was a significant factor in determining the customer satisfaction (McDougall & Levesque, 2000). Hence, it is suggested that perceived value can affect tourists' trip satisfaction.

Hypothesis 9. Tourist's perceived value of the trip positively affects tourist's travel satisfaction.

Additionally, McDougall and Levesque (2000) have proposed a model in which perceived value has a direct impact on the customer's satisfaction and then results in future behavioral intentions. Also, Bolton and Drew (1991) argue that customer's future behavioral intention is determined by some variables which perceived value is an important factor among them. On the other hand, customers will be motivated to buy a service or product when they evaluate whether they have received value for the service or not (McDougall & Levesque, 2000). Therefore, it can be argued that a tourist's perceived value can influence his/her future behavioral intentions to visit a destination.

Hypothesis 10. Tourist's perceived value of the trip positively affects tourist's behavioral intention to visit the destination.

2.6 Satisfaction

One of the main goals of the businesses is to focus on achieving customer satisfaction (Woodruff, 1997). This factor becomes even more important for the service firms (McDougall & Levesque, 2000). Understanding about satisfaction provides useful information for the managers which help them to make decisions more effectively (Danaher & Haddrell, 1996). Moreover, satisfaction can be considered as an outcome of the customer evaluation and assessment about a product or service (Feng, 2007).

Chen and Tsai (2007) have proposed a model in which they show that customer satisfaction affect the behavioral intention. In their study the results show that when the higher level of satisfaction would occur, the more positive intention could brought out (Chen & Tsai, 2007).

In addition a study has been conducted in Sri Lanka that examines the relationship between tourists 'satisfaction and behavioral intention. The results show that improving in the customer satisfaction element creates positive behavioral intention toward this country (Ranasinghe, 2012).

Hypothesis 11. Tourist's overall travel satisfaction positively affects tourist's behavioral intention to visit the destination.

Satisfaction is also considered as a direct antecedent of travelers' loyalty and intention to revisit a destination especially in the short period of time (Feng, 2007). In other words, satisfaction has a direct, positive and significant impact on the immediate revisit intention (Assaker et al., 2011). Yang and Peterson (2004) state that according to the web-based survey among the online service users of their sample, in order to achieve customer loyalty firms need to focus on the satisfaction as an important objective (Yang & Peterson, 2004). Moreover, Bloemer and Kasper (1995) argued that travelers evaluate the performance of the firm and then compare with their expectations. When the performance is higher than expectation, satisfaction will occur which results in the customer loyalty (Bloemer & Kasper, 1995).

Hypothesis 12. Tourist's overall travel satisfaction positively affects tourist's destination loyalty.

2.7 Behavioral intention

Behavioral intention is a desire to visit or revisit a destination. Thus, when this desire transforms into the action, customer loyalty will occur (Yang & Peterson, 2004). Oliver (1999) has proposed 4 stages for customer loyalty. The first stage is cognitive loyalty meaning that the traveler becomes loyal to the destination when he/she has enough information about the destination. Second stage is affective loyalty which means having a positive attitude toward a special destination. The third stage is conative loyalty or behavioral intention and the last stage is action loyalty. In this study customer loyalty is directly measured by the customer behavioral intention (Oliver, 1999). Hence, understanding the behavioral intention is very critical in understanding and predicting the future intention of travelers and their loyalty toward a special destination (Assaker et al., 2011).

Hypothesis 13. Tourist's behavioral intention to visit the destination positively affects tourist's destination loyalty.

2.8 Loyalty

It is widely accepted that loyalty is a major driving force for the success of businesses. The academic literature is replete with definitions of loyalty. For example, in a traditional approach, loyalty depends on customers' repeated purchase of a product or service (e.g. Cunningham, 1956). However, it is suggested that loyalty is a complicated and multifaceted phenomenon. For example, Dick and Basu (1994) propose an integrated loyalty framework that portrays loyalty as a relationship between the relative attitude toward an entity (e.g. a brand) and patronage behaviour. Moreover, in an attitudinal view, customer loyalty depends on a continuous relationship with the service provider (Yang & Peterson, 2004). In a tourism context loyalty refers to tourists' revisit of a destination and their positive word-of-mouth recommendations (Chi & Qu, 2008). Through a behavioral lens, it is suggested that tourists' past experience affects their future travel choices and more specifically, their destination choice (Oppermann, 2000). According to Woodside and MacDonald (1994), based on destination choice history, one can divide tourists into two groups: continuous repeaters and continuous switchers, or in other words undivided loyalty and no loyalty (Brown, 1953).

3. RESEARCH METHODOLOGY

3.1 Research design

The aim of this study is to look at the possible relationships between factors including food image, destination image, information source, trip quality, perceived value, satisfaction, behavioral intention and loyalty. For this purpose, a quantitative approach was selected to get more insights towards the very depth of this matter. Malhotra (2008) argues that the major objective using this design is to gain deeper insights of the issues or the ideas from the respondents.

3.2 Sampling

Malaysia was selected as the destination for the purpose of this research's objectives. The reason for this choice is that, the population in Malaysia consist of various ethnics and cultures and the ethnic diversity is obvious in this country (Ling et al., 2010). In addition Malaysia is among the top 10 destinations by arrivals in the world (UNWTO, 2013). Therefore, the target population for this study was tourists that had taken a trip to Malaysia in the previous 12 months. Regarding sample size, a sample size between 200 and 400 is normally recommended and accepted as a critical sample size (Hair, 2010). Therefore a target sample size of 200 respondents were selected which were from various countries with different backgrounds.

A convenience sampling method was chosen as the participants were selected based upon their availability and willingness to participate and also, because it was the most feasible method given the budget and time constraints. The study was carried out in Malaysia. Firstly, The Juta Mines Condominium and South City in Serdang was chosen. The reason was that these places are full of foreign residents and potential tourists who mostly come from Asian countries such as Iraq and Iran. Then, KLIA airport, Ampang Street and the area around KLCC such as Twin Towers were chosen. The reason to choose such areas was that these areas are among the most famous tourist attractions in Kuala Lumpur which are full of tourists and were suitable to get the 200 samples needed to conduct this study. Next, every third level in apartments and every tenth person in the open areas was chosen to distribute the questionnaire. In addition, an online version of the questionnaire was designed. An invitation letter explaining research's objective and containing a link to the questionnaire was sent to emails of the National University of Malaysia (UKM) students as it is one of the most international-friendly universities in Malaysia. Moreover, the travelers in the area of KLCC were asked to give their emails to send the online questionnaire to them. The results of the survey was stored in a Microsoft Excel file.

3.3 Research instrument

A self-administrated questionnaire was used as a means to achieve the objectives of this study. In this questionnaire, a six-point Likert scale was utilized since, as noted by Garland (1991), in the questionnaires which do not have the mid-point on a Likert scale, the respondents have no chance to answer merely to please the interviewer or to hide their real perceptions and opinions which are socially unacceptable. In fact, by eliminating the mid-point, all these issues mentioned before will be minimized.

The questions in the questionnaire are based on a review of the literature and specific characteristics of Malaysia destination. The questionnaire consisted of eight sections. Included categories of questions were food image, destination image, information source, trip quality, perceived value, satisfaction, behavioral intention and loyalty. Destination image and trip quality were consisted of subsections.

3.4 Measures

The questionnaire items used in this study to measure the variables were taken from previous studies to ensure the validity of research. Table 1 summarizes the sources and Cronbach's alpha for each construct used in this study.

Table 1. Constructs of the study

Factor	Source	Cronbach's alpha
Food Image (FI)	(Ling et al., 2010)	Higher than 0.8
Destination Image (DI)	(Chen & Tsai, 2007)	Higher than 0.8
Information Sources (IS)	(Ab Karim, 2006)	Higher than 0.8
Trip Quality (TQ)	(Chen & Tsai, 2007)	Higher than 0.8
Perceived Value (PV)	(Chen & Tsai, 2007)	0.91
Satisfaction (SAT)	(Chi & Qu, 2008)	0.91
Behavioral Intention (BI)	(Alam et al., 2012)	0.97
Loyalty	(Roodurmun, 2010)	0.87

Part 1 of the questionnaire dealt with the measurement of food image with food attributes. Part 2 contained variables which measured the destination image and it was divided into 3 subsections which were destination brand, entertainment and the nature and culture. The third part dealt with the measurement of information sources. Part 4 included variables to measure the trip quality and it had 4 subsections including hospitality, attractions, transportation and amenity. The fifth part was designed to measure the perceived value. Part 6 of the questionnaire dealt with the calculation of satisfaction. The seventh part measured the behavioral intention and the next part assessed the loyalty of the travelers. Lastly, demographic characteristics of respondents were collected.

3.5 Data analysis

Collected data was coded and entered into SPSS software. The analysis was carried out in three consecutive stages. First a descriptive analysis was performed to explain the state of food image, destination image, information sources, trip quality, perceived value, satisfaction, behavioral intention and loyalty. Furthermore, demographic characteristics of respondents were analyzed using a descriptive approach through frequency tables. Second, factor analysis was utilized to group the variables into common themes. The type of factor analysis was principal component analysis and varimax was the rotation technique in this study. Finally, multiple regression analysis was conducted to test the proposed hypotheses and to determine the relationships between independent and dependent variables of the conceptual model.

4. DISCUSSIONS OF THE FINDINGS

4.1 Respondents' demographic profile

Table 2 reports the personal and demographic characteristics of the respondents. Approximately 60% of the respondents were male. Most of the respondents (N=89; 44%) were between 30 to 39 years old. More than 75% of the respondents were graduated or post graduated. Respondents were from 23 different countries and over 25% of the respondents were Iranian (N=43; 23.1%) and Iraqi (N=28; 13.9%)

people. Chinese (N=20; 9.9%) and Swedish (N=20; 9.9%) were the next popular tourist nationalities of the study's sample.

4.2 Reliability

As Hair (2010) argues, reliability is concerned about the consistency of the measurements. The internal reliability of the items was confirmed by computing the Cronbach's alpha which is considered as an accepted measure of reliability (see Table 3). The Cronbach's alpha for all factors were above 0.7, which is considered as an acceptable cut-off value (Hair, 2010), meaning that all factors in this study were reliable.

4.3 Content validity

Validity refers to the degree to which a set of measures represent the concept of a study. In fact validity means how well the concept is defined by a set of measures (Hair, 2010). To ensure the validity of research, the questionnaire is based both on previous studies (see Table 1) as well as on the opinions of expert in the area of tourism and marketing.

4.4 Factor analysis

Factor analysis is a method by which a simple structure or pattern of results will be generated so that each variable load highly onto one factor and can help the researcher to reach the objectives of either data summarization or data reduction. Rotated Factor Matrix shows the correlations between the variable and the factor (see Table 3). Correlation of 0.4 or less is a low correlation and is considered as insignificant. In this analysis, variances with loadings with the minimum value of 0.4 were considered in the further analysis. The sampling adequacy of the data used in this study was checked with KMO measure and was determined to be 0.903 which is considered as an excellent sampling adequacy (Kaiser, 1974). In addition, Bartlett's Test of Sphericity (BTS) was significant in this study. Furthermore, cumulative variance explained was 65.46% in this study. Despite the fact that no absolute threshold has been adopted, a 60% cumulative variance is quite commonly accepted in social sciences (Hair, 2010).

Table 2. Respondent profile

Demographic characteristics	Frequency	Percentage (%)
1. Gender		
Female	82	40.6
Male	120	59.4
2. Age group		
Less than 20 years old	6	3.0
20-29 years old	77	38.1
30-39 years old	89	44.1
40-49 years old	27	13.4
50-59 years old	3	1.5
60 years old and older	0	0.0
3. Marital status		
Married	104	51.5
Never married	90	44.6
Divorced/Widowed/Separated	8	4.0
4. Highest education		
Some high school or less	1	0.5
High school graduate	9	4.5
College/University graduate	37	18.4
Graduate/Post-graduate	155	76.6

4.5 Hypothesis testing

4.5.1 Multicollinearity

Hair (2010) argues that variables which are highly collinear can distort the results and such results cannot be generalized. Two common measures to assess multicollinearity are the tolerance and variance inflation factor (VIF) values (see Table 4). Tolerance values should be above 0.10 and the value of VIF should be below 5. In the current study, the tolerance values of all of the variables are above 0.10. Likewise, the VIF value is less than 5, thus further confirming that the multicollinearity problem is not a concern in this study. The acceptable Durbin-Watson range is between 1.5 and 2.5. In this analysis all 6 extracted Durbin-Watsons values were between the acceptable ranges, proving that there were no autocorrelation problems in the data used in this research. Thus, the measures selected to assess independent variables in this study do not reach levels indicative of multicollinearity.

4.5.2 Multiple regression analysis

This section is devoted to presenting the results of hypothesis testing. To test the hypotheses, multiple regression analysis was utilized according to guidelines proposed by Hair (2010). The first step in regression analysis of this study, concerned respectively Hypothesis 1 (i.e. Food image of the destination positively affects tourist's trip quality), Hypothesis 3 (i.e. Destination image positively affects tourist's trip quality) and Hypothesis 5 (i.e. The type of information source has a significant effect toward the tourist's trip quality). As such, the aim was to test the relationship between the independent variables including food image, destination image and information sources, and the dependent variable of trip quality. As it is demonstrated in Table 5, all of the hypotheses were found to be significant in the prediction model. Furthermore, the results support the hypotheses; that is, the relationships among food image (β=0.286; p of 1.576 in this relationship falls within the acceptable range which is between 1.5 and 2.5, suggesting that there is no auto-correlation problems in the data used in this study.

The second step in regression analysis of this study concerned respectively Hypothesis 2 (i.e. Food image of the destination positively affects tourist's perceived trip value), Hypothesis 4 (i.e. Destination image positively affects tourist's perceived trip value) and Hypothesis 6 (i.e. The type of information source has a significant effect toward the tourist's perceived trip value). As such, the aim was to test the relationship between the independent variables including food image, destination image and information sources, and the dependent variable of perceived value. As it is demonstrated in Table 6, Hypothesis 4 was found to be significant whereas the Hypotheses 2 and 6 were not supported in the prediction model. The results of regression analysis support the hypothesis of destination image (β=0.678; p0.05) and information sources ($p>0.05$) as predictors of perceived value. Also, the DurbinWatson value of 1.775 in this relationship falls within the acceptable range which is between 1.5 and 2.5, suggesting that there are no auto-correlation problems in the data used in this study

In the third part of the regression analysis of this study, Hypothesis 7 (i.e. Trip quality positively affects tourist's travel satisfaction) and Hypothesis 9 (i.e. Tourist's perceived value of the trip positively affects tourist's travel satisfaction) were tested. The goal was to test the relationship between the independent variables including trip quality and perceived value, and the dependent variable which was satisfaction. As it is demonstrated in Table 7, all of the hypotheses were found to be significant in the prediction model. Furthermore, the results support the hypotheses; that is, the relationships among trip quality ($\beta=0.467$; p

Table 3. Rotated component matrix, reliability statistics and mean score

Factor	Variable	Loading	Eigenvalue	Cronbach's Alpha
Food image	Food in Malaysia has reasonable price	0.541	2.844	0.878
	Malaysia has high availability of food	0.410		
	Food providers in Malaysia are friendly	0.464		
	Malaysian foods are appealing to human senses	0.581		
	Services by food providers are good in Malaysia	0.580		
	Food spots are located in convenient places	0.561		
	Malaysian foods are clean and safe to consume	0.616		
	Foods are well presented in Malaysia	0.755		
	Malaysian chefs use variety of cooking methods	0.546		
	Malaysian restaurants have comfortable eating surroundings	0.605		
Destination image	Malaysia offers personal safety	0.431	25.571	0.901
	Malaysia has a good quality of life	0.432		
	Malaysia is a clean destination	0.423		
	Malaysia has a good name and reputation	.419		
	Malaysian people are hospitable and friendly	0.433		
	Malaysia has a good nightlife	0.613		
	Malaysia is a good place for shopping	0.766		
	Malaysia offers varied food choices	0.688		
	Malaysia offers exotic food choices	0.663		
	Malaysia has a great variety of fauna and flora	0.678		
	Malaysia has a spectacular landscape	0.774		
	Malaysia has its unique ways of life and customs	0.634		
Information sources	Brochure/Travel guides are important information sources for me	0.797	3.972	0.922
	Travel book is an important information source for me	0.865		
	Travel magazines are important information sources for me	0.848		
	Highway welcome center is an important information source for me	0.765		
	Local/National tourism offices are important	0.711		

Factor	Variable	Loading	Eigenvalue	Cronbach's Alpha
	information sources for me			
	Newspapers are important information sources for me	0.618		
	Direct mail from destination is an important information source for me	0.749		
	Travel agency/Company are important information sources for me	0.733		
	TV/Movies/Travel channels are important information sources for me	0.640		
Trip quality	Price of accommodation is reasonable in Malaysia	0.685	2.318	0.890
	Price of activities is acceptable in Malaysia	0.540		
	Food and beverage of accommodation is acceptable in Malaysia	0.513		
	Prices of food and beverage are reasonable in Malaysia	0.607		
	Cleanness of beaches is acceptable in Malaysia	0.429		
	The landscape is very unique in Malaysia	0.440		
	Malaysia has beautiful beaches	0.420		
	Transportation is easy to access	0.658		
	Internal transport are available everywhere	0.748		
	Parking facilities and spaces are enough	0.697		
	General infrastructure is acceptable	0.610		
	Travel information is easily accessible	0.587		
Perceived value	My trip to Malaysia worth the money I spent for the trip	0.419	25.571	0.916
	My trip to Malaysia worth the effort I made for the trip	0.516		
Satisfaction	Attractions in Malaysia satisfied me	0.516	2.164	0.864
	Shopping in Malaysia satisfied me	0.661		
	Dining in Malaysia satisfied me	0.653		
	Activities and events in Malaysia satisfied me	0.562		
Behavioral intention	It is likely that I visit Malaysia	0.406	1.380	0.892
	I expect to visit Malaysia in the future	0.735		
	I predict I will visit Malaysia	0.799		
	I plan to visit Malaysia as my holiday destination	0.734		
Loyalty	It's very important for me to visit Malaysia more than the other countries	0.641	5.976	0.898
	I always visit Malaysia because I really like this country	0.711		
	I'm going to visit Malaysia in the future	0.764		
	I'm going to remain loyal to Malaysia as tourist destination	0.716		

KMO=0.903, Sig. of BTS=0.000, Cum. Var. explained (%)=65.464%, principal component, varimax rotation.

The fourth section of the regression analysis of this study aims at testing Hypothesis 8 (i.e. Trip quality positively affects tourist's behavioral intention to visit the destination) and Hypothesis 10 (i.e. Tourist's perceived value of the trip positively affects tourist's behavioral intention to visit the destination). The goal was to test the relationship between the independent variables including trip quality and perceived value, and the dependent variable which was behavioral intention. As it is shown in Table 8, all of the hypotheses were found to be significant in the prediction model. Furthermore, the results support the hypotheses; that is, the relationships among trip quality ($\beta=0.510$; $p<0.05$) perceived value ($\beta=0.288$; $p<0.05$). The result show that both trip quality and perceived value affect behavioral intention. In addition the beta coefficients show that the trip quality impacts on behavioral intention more significantly compare to the other factor which is perceived value. Also, the Durbin-Watson value of 2.052 in this relationship falls within the acceptable range which is between 1.5 and 2.5, suggesting that there are no autocorrelation problems in the data used in this study

Table 4. Test of collinearity

Factor	Tolerance	VIF
Food image	0.626	1.597
Destination image	0.608	1.645
Information sources	0.944	1.059
Trip quality	0.582	1.719
Perceived value	0.582	1.719
Satisfaction	0.612	1.635
Behavioral intention	0.612	1.635

Table 5. Regression results, Dependent variable: Trip quality

Factor	Standardised beta coefficient	T	Sig.
(Constant)		1.766	0.079
Food Image	0.286	4.863	0.000
Destination Image	0.505	8.465	0.000
Information Source	0.130	2.715	0.007

Durbin-Watson=1.576, F=88.291, R^2=0.572, Adj. R^2=0.566, Dependent variable: Trip quality

In the fifth part of the regression analysis of this study, Hypothesis 11 (i.e. Tourist's overall travel satisfaction positively affects tourist's behavioral intention to visit the destination) was tested. The aim was to test the relationship between the independent variable which was satisfaction, and the dependent variable which was behavioral intention. As it is demonstrated in Table 9, this hypothesis was found to be significant in the prediction model. Furthermore, the results support the hypothesis. The beta coefficients ($\beta=0.623$; p

Table 6. Regression results, Dependent variable: Perceived value

Factor	Standardised beta coefficient	T	Sig.
(Constant)		-1.919	0.056
Food Image	0.084	1.390	0.166
Destination Image	0.678	11.082	0.000
Information Source	0.035	0.706	0.481

Durbin-Watson=1.775, F=80.461, R^2=0.549, Adj. R^2=0.543, Dependent variable: Perceived value

Table 7. Regression results, Dependent variable: Satisfaction

Factor	Standardised beta coefficient	T	Sig.
(Constant)		5.751	0.000
Trip Quality	0.467	7.183	0.000
Perceived Value	0.317	4.880	0.000

Durbin-Watson=2.064, F=103.782, R^2=0.511, Adj. R^2=0.506, Dependent variable: Satisfaction

Table 8. Regression results, Dependent variable: Behavioral intention

Factor	Standardised beta coefficient	T	Sig.
(Constant)		-0.153	0.879
Trip Quality	0.510	8.038	0.000
Perceived Value	0.288	4.536	0.000

Durbin-Watson=2.052, F=113.754, R^2=0.533, Adj. R^2=0.529, Dependent variable: Behavioural intention

Table 9. Regression results, Dependent variable: Behavioral intention

Factor	Standardised beta coefficient	T	Sig.
(Constant)		1.703	0.090
Satisfaction	0.623	11.266	0.000

Durbin-Watson=2.018, F=126.914, R^2=0.388, Adj. R^2=0.385, Dependent variable: Behavioural intention

The last section of the regression analysis of this study aimed at testing Hypothesis 12 (i.e. Tourist's overall travel satisfaction positively affects tourist's destination loyalty) and Hypothesis 13 (i.e. Tourist's behavioral intention to visit the destination positively affects tourist's destination loyalty). The goal was to test the relationship between the independent variables including satisfaction and behavioral intention, and the dependent variable which was loyalty. As it is demonstrated in Table 7, while Hypothesis 13 was significant, Hypothesis 12 was not supported in this study. That is to say, the results support the hypothesis; that is, the relationship between behavioral intention (β=0.844; p0.05) hypothesis was rejected in this study. Also, the Durbin-Watson value of 2.147 in this relationship falls within the acceptable range which is between 1.5 and 2.5, suggesting that there is no auto-correlation problems in the data used in this study.

Table 10. Regression results, Dependent variable: Loyalty

Factor	Standardised beta coefficient	T	Sig.
(Constant)		-1.450	0.149
Satisfaction	0.028	0.599	0.550
Behavioural Intention	0.844	18.332	0.000

Durbin-Watson=2.147, F=286.151, R^2=0.742, Adj. R^2=0.739, Dependent variable: Loyalty

Table 11 summarizes the results of hypothesis testing of this study. To sum up, only Hypothesis 2, Hypothesis 6 and Hypothesis 12 were not supported while all other hypotheses were supported.

5. DISCUSSIONS

5.1 Food image

As results show, food image is a factor which influences trip quality while it is not a significant predictor of perceived value. Previous studies found that food image has significant positive impact on trip quality (Chen & Tsai 2007; Ab Karim 2006). Some studies (e.g. Good et al., 2008; Kuo et al., 2009; Bauer et al., 2006) show that food image affects service quality and then service quality affects perceived value. Therefore, it might be stated that food image does not have a direct impact on perceived value. In addition multiple regression analysis which shows the results of food image in relation with trip quality

(β=0.286, p-value= 0.000) and in relation with perceived value (β=0.084, p-value=0.166) points out that the food image is a factor which affects the trip quality while it has no impact on the perceived value in this study. It seems that a positive perception feeling of the travelers toward food in a certain destination can increase the trip quality for the travelers while this important factor does not play an important role in the improving the travelers' perceived trip value.

Table 11. Summary of hypothesis testing

Hypothesis	Testing result
H1. Food image of the destination positively affects tourist's trip quality	Supported
H2. Food image of the destination positively affects tourist's perceived trip value.	Rejected
H3. Destination image positively affects tourist's trip quality	Supported
H4. Destination image positively affects tourist's perceived trip value	Supported
H5. The type of information source has a significant effect toward the tourist's trip quality.	Supported
H6. The type of information source has a significant effect toward the tourist's perceived trip value.	Rejected
H7. Trip quality positively affects tourist's travel satisfaction.	Supported
H8. Trip quality positively affects tourist's behavioural intention to visit the destination	Supported
H9. Tourist's perceived value of the trip positively affects tourist's travel satisfaction	Supported
H10. Tourist's perceived value of the trip positively affects tourist's behavioural intention to visit the destination	Supported
H11. Tourist's overall travel satisfaction positively affects tourist's behavioural intention to visit the destination	Supported
H12. Tourist's overall travel satisfaction positively affects tourist's destination loyalty	Rejected
H13. Tourist's behavioural intention to visit the destination positively affects tourist's destination loyalty	Supported

5.2 Destination image

Higher level of travelers' knowledge and feelings toward the destination is associated with the increase in both trip quality and perceived value. Studies like Chen and Tsai (2007), state that when a destination image is favorable for the traveler, the result is higher trip quality and the higher perceived value. Regression analysis confirms these findings and shows results of destination image in relation with trip quality (β=0.505, p-value= 0.000) and in relation with perceived value (β=0.678, p-value=0.000) indicates that destination image has positive impact on both trip quality and perceived value.

5.3 Information sources

The findings of this study clearly show that information sources have a positive relationship with trip quality. The regression analysis shows the results of information source in relation with trip quality (beta=0.130, p-value=0.007) confirm the positive impact of the information source on the trip quality. This is in line with the arguments of McIntosh et al. (1995) and Snepenger et al. (1990) who argue travelers who try to catch a higher level of information and knowledge about a trip would enjoy higher trip quality. On the other hand the regression analysis results of information source in relation with perceived value (beta=0.035, p-value=0.481>0.05) indicates that the relationship between information source and perceived value is not significant. The reason could be that some hidden variables such as gender moderate the relationship between information sources and travelers' perceived value. As such, future studies might address this issue by analyzing the possible moderators in the abovementioned relationship.

5.4 Trip quality

Multiple regression result points out that trip quality is a significant and positive predictor of both satisfaction and the behavioral intention. Regression analysis shows results of trip quality in relation with satisfaction (β=0.467, p-value= 0.000) and in relation with behavioral intention (β=0.510, p-value=0.000) indicates that trip quality positively affects satisfaction and behavioral intention of the travelers toward a destination. Likewise, Žabkar et al. (2010) and Murphy et al. (2000) state that when travelers have positive feeling toward their trip, the quality of the trip will exceed and then results in the increased of travelers' satisfaction and behavioral intention toward the trip.

5.5 Perceived value

Perceived value is a factor which has a positive impact on the improvement of travelers' satisfaction and behavioral intention in tourism industry. Regression analysis results of perceived value in relation with satisfaction (β=0.317, p-value= 0.000) and in relation with behavioral intention (β=0.288, p-value=0.000) showed that the more positive evaluation of the travelers of the value of their trip, the more satisfied they will be and the more intention they will have to visit a destination. This is consistent with the study done by McDougall and Levesque (2000) and Bolton and Drew (1991), in which perceived value has a direct impact on the customer's satisfaction and future intentions.

5.6 Satisfaction

The higher level of satisfaction is associated with increased future intention to visit a destination. Multiple regression analysis shows the results of satisfaction (β=0.623, p-value= 0.000) implying that there is a positive and significant correlation between satisfaction and behavioral intention. The current study therefore, further proves the earlier findings that showed when the higher level of satisfaction would occur, the more positive intention could brought out (Ranasinghe, 2012; Chen & Tsai, 2007). The point to make here is that satisfaction is not significantly associated with increase in travelers' loyalty. The regression results of satisfaction (β=0.028, p-value=0.550>0.05) confirms this claim. The amount of p-value in the regression results of satisfaction indicates that the relationship between satisfaction and loyalty is not significant. This contradicts the results of studies done by Assaker et al. (2011); Yang and Peterson (2004) and Bloemer and Kasper (1995) who argue tourists satisfaction of their travel lead to their loyalty toward the destination. The special characteristics of this study's sample in destination Malaysia might be one reason why this relationship was failed to be confirmed.

5.7 Behavioral intention

Tourist's behavioral intention to visit the destination positively and significantly affects tourist's destination loyalty. Studies like those carried out by Assaker et al. (2011); Oliver (1999) and Yang and Peterson (2004) have generally shown that behavioral intention is a desire to visit or revisit a destination which will transform into the action and results is customer loyalty. As shown in Table 10, regression analysis shows the results of behavioral intention (β=0.844, pvalue= 0.000) indicating that the intention to visit or revisit a destination significantly leads to the travelers' loyalty toward the destination. Therefore the results of this study prove the earlier findings which were mentioned above.

6. IMPLICATIONS

6.1 Implications for research

The current study attempted to propose an integrated framework for examining the antecedents of loyalty intention to travel to Malaysia destination. It would be appropriate to explore more comprehensive models for the drivers of loyalty and future research can include more factors such as the role of a destination's authenticity in perceived travel value in the model to make it even more integrated. Extending this research to incorporate other measures and broader definitions would offer further insights and understanding of how destinations can increase loyalty among travelers. Moreover identifying the possible moderators of the research model (e.g. nationality, marital status, education) and analyzing their effect on related constructs can be developed by future research. In addition, using larger sample, more different destinations and also using a random technique for sampling that will reduce the possible bias as much as possible is suggested. Future researchers could also address the issue of time, as for example tourists generally tend to travel more in summer, and therefore the researchers could conduct similar surveys in different seasons to collect relevant respondents and make more comprehensive surveys. Furthermore this study concentrated on only one consequence of the model which was travelers' loyalty to a destination. Other consequences of all constructs used in this research have to be studied in other research. Last but not least, the sample used in this study focused on literate visitors with the ability of reading the English language. It might be important to develop ways of measuring/investigating the influence of all factors used in the current study on travelers' loyalty in different languages.

6.2 Implication for practice

This study's findings have significant managerial implications for destination managers and marketers in various organizations such as Tourism Malaysia, Kuala Lumpur City Hall (Dewan Bandaraya Kuala Lumpur, DBKL) and the Ministry of Tourism and Culture Malaysia. As Assaker et al. (2011) argue, in today's increasingly competitive travel market, it is very difficult for marketers to attract more travelers. Tourists can choose among many different destination options and enjoy their trip. Hence it is very critical for destinations to focus on travelers' intention to revisit and not just actual visits (Assaker et al., 2011). Thus, to help managers achieve a more in-depth understanding of tourists' intention to revisit the destination the present study applies various constructs, which facilitates interpretation and enables a more accurate prediction of travelers' loyalty to Malaysia. The results show the importance of food image, destination image, information source, trip quality, perceived value, satisfaction and behavioral intention on travelers' willingness to revisit the destination. Hence tourism industries should use more sophisticated strategies such as establishing strategic events and ceremonies to grab more tourists. It is worth noting that residents can play an important role to increase the trip quality of the travelers. Therefore managers might need to use different strategies in order to inform the local residents about the importance of tourism industry and its significant impact on the economy.

The findings of this study could also help destinations to formulate the type of food image they want to establish and communicate. Managers therefore need to focus on using food as a branding tool in terms of brand differentiation and destination image.

As there was a statistically significant relationship between tourists' satisfaction and behavioral intention, it is strongly recommended that managers should improve the travelers' satisfaction by using strategies in which the trip quality will be improved in order to increase satisfaction toward destination. Particularly, government should take necessary actions to improve the infrastructure facilities such as increasing the availability of public transportation since the item regarding the availability of internal transport scored one of the lowest means (mean=3.50) among items of this study.

7. CONCLUSIONS

The purpose of this study was to examine the factors affecting travellers' loyalty towards Malaysia as a tourism destination. The study contributes to the existing literature of tourism by providing an integrated framework of loyalty antecedents and widens our understanding of travellers' intention to revisit as a critical factor in tourism industry, while also examines the rationales for intention or reluctance to revisit the Malaysia. The study also offers some interesting managerial insights. As such, the findings of the current research provide support for both investment decisions, and decisions relating to the development of tourism industry that address their needs into consideration.

The theoretical framework of the present study was developed based on previous studies. The multiple regression analysis that was utilized in this study showed that food image, destination image and information source are critical factors which impact trip quality while among these factors just destination image affects the perceived trip value. Furthermore both trip quality and perceived trip value are very important elements of tourists' satisfaction and behavioural intention. In addition, satisfaction is very important element of behavioural intention; however, the impact of travellers' satisfaction on intention to revisit a destination was not statistically significant in the present study. It should be emphasized that behavioural intention is an important factor which affects tourists' intention to revisit a destination. The model explains 65% of the variance in tourists' intention to revisit a destination. As the Malaysia tourism industry is constantly growing (Aruna, 2013), an understanding of the factors that influence the travellers' intention to revisit Malaysia becomes more important.

In spite of the study's limitations, this research makes some important contributions. The existing literature reviewed in the area of tourism and the theoretical framework was developed. Furthermore this study provides the analysing of factors which affect intention to revisit a destination which improves the tourism field as a whole. Last but not least, the model which was developed in this study has not been studied to date and can be a starting point for developing integrated models of loyalty in a tourism context.

8. LIMITATIONS AND FUTURE RESEARCH

Despite the implications generated from this research, like other studies this study has some limitations as well. The sample used for this research was consisted of travelers in Kuala Lumpur, which might limit the generalizability of the results. Future research need to focus on the more diversified random samples to verify the findings of the current study and use participants in other geographic areas. Moreover if the current study could provide a larger sample population, the results would be more accurate. In addition, it is completely obvious that the respondents are different in terms of visiting different cities in Malaysia. Hence future research could compare the difference between the visitors who have visited some cities in Malaysia with those who have just visited Kuala Lumpur. Last but not least, the study could use more variables to find more results and findings. However it should be noted that a long questionnaire might discourage the respondents to participate in the survey.

REFERENCES

1. AB KARIM, S. 2006. Culinary tourism as a destination attraction: An empirical examination of the destination's food image and information sources. Oklahoma State University.

2. ALAM, S. S., JANO, Z., WEL, C., ANIZA, C. & AHSAN, N. 2012. Is Religiosity an Important Factor in influencing the intention to undertake Islamic Home Financing in the Klang Valley? *World Applied Sciences Journal*, 19, 1030-1041.

3. ARUNA, P. 2013. Giving the tourism sector a boost [Online]. The star online. Available: http://www.thestar.com.my/news/nation/2013/08/15/giving-the-tourism-sector-a-boost.aspx/ [Accessed 18 June 2014].

4. ASSAKER, G., VINZI, V. E. & O'CONNOR, P. 2011. Examining the effect of novelty seeking, satisfaction, and destination image on tourists' return pattern: A two factor, non-linear latent growth model. *Tourism Management*, 32, 890-901.

5. BADARNEH, M. B. & SOM, A. P. M. 2011. Factors Influencing Tourists' Revisit Behavioral Intentions and Loyalty.

6. BAKER, M. S. P. 2014. Assessment of the sustainability of Little Kulala Camp and Kulala Wilderness Camp in Namibia.

7. BAUER, H. H., FALK, T. & HAMMERSCHMIDT, M. 2006. eTransQual: A transaction process-based approach for capturing service quality in online shopping. *Journal of Business Research*, 59, 866-875.

8. BLOEMER, J. M. & KASPER, H. D. 1995. The complex relationship between consumer satisfaction and brand loyalty. *Journal of economic psychology*, 16, 311-329.

9. BOLTON, R. N. & DREW, J. H. 1991. A longitudinal analysis of the impact of service changes on customer attitudes. *Journal of marketing*, 55.

10. BOLTON, R. N. & LEMON, K. N. 1999. A Dynamic Model of Customers' Usage of Services: Usage as an Antecedent and Consequence of Satisfaction. *Journal of Marketing Research* (JMR), 36.

11. BROWN, G. H. 1953. Brand Loyalty-fact of fiction. Trademark Rep., 43, 251.

12. CHEN, C.-F. & TSAI, D. 2007. How destination image and evaluative factors affect behavioral intentions? Tourism management, 28, 1115-1122.

13. CHI, C. G.-Q. & QU, H. 2008. Examining the structural relationships of destination image, tourist satisfaction and destination loyalty: An integrated approach. Tourism management, 29, 624-636.

14. COLE, S. T. & ILLUM, S. F. 2006. Examining the mediating role of festival visitors' satisfaction in the relationship between service quality and behavioral intentions. *Journal of Vacation Marketing*, 12, 160-173.

15. CRONIN JR, J. J., BRADY, M. K. & HULT, G. T. M. 2000. Assessing the effects of quality, value, and customer satisfaction on consumer behavioral intentions in service environments. *Journal of retailing*, 76, 193-218.

16. CUNNINGHAM, R. M. 1956. Brand loyalty-what, where, how much. Harvard Business Review, 34, 116-128.

17. DANAHER, P. J. & HADDRELL, V. 1996. A comparison of question scales used for measuring customer satisfaction. *International Journal of Service Industry Management*, 7, 4-26.

18. DICK, A. S. & BASU, K. 1994. Customer loyalty: toward an integrated conceptual framework. *Journal of the academy of marketing science*, 22, 99-113.

19. FENG, R. 2007. Temporal destination revisit intention: The effects of novelty seeking and satisfaction. Tourism management, 28, 580-590.

20. GARLAND, R. 1991. The mid-point on a rating scale: Is it desirable. Marketing Bulletin, 2, 66- 70.

21. GOOD, D. H., SICKLES, R. C. & WEIHER, J. C. 2008. A hedonic price index for airline travel. Review of Income and Wealth, 54, 438-465.

22. GRUEN, T. W., OSMONBEKOV, T. & CZAPLEWSKI, A. J. 2006. eWOM: The impact of customer-to-customer online know-how exchange on customer value and loyalty. Journal of Business research, 59, 449-456.

23. HAIR, J. F. 2010. Multivariate Data Analysis, Prentice Hall.

24. HJALAGER, A. M. & CORIGLIANO, M. A. 2000. Food for tourists—determinants of an image. International Journal of Tourism Research, 2, 281-293.

25. KAISER, H. F. 1974. An index of factorial simplicity. Psychometrika, 39, 31-36.

26. KUO, Y.-F., WU, C.-M. & DENG, W.-J. 2009. The relationships among service quality, perceived value, customer satisfaction, and post-purchase intention in mobile value-added services. Computers in human behavior, 25, 887-896.

27. LIN, Y.-C. 2006. Food images in destination marketing, ProQuest.

28. LING, L. Q., KARIM, M. S. A., OTHMAN, M., ADZAHAN, N. M. & RAMACHANDRAN, S. 2010. Relationships Between Malaysian Food Image, Tourist Satisfaction and Behavioural Intention. *World Applied Sciences Journal*, 10, 164-171.

29. LIU, C.-H., TZENG, G.-H. & LEE, M.-H. 2012. Improving tourism policy implementation–the use of hybrid MCDM models. Tourism Management, 33, 413-426.

30. LONG, L. M. 2013. Culinary tourism, University Press of Kentucky.

31. MALHOTRA, N. K. 2008. Marketing Research: An Applied Orientation, 5/E, Pearson Education India.

32. MCDOUGALL, G. H. & LEVESQUE, T. 2000. Customer satisfaction with services: putting perceived value into the equation. Journal of services marketing, 14, 392-410.

33. MCINTOSH, R. W., GOELDNER, C. R. & RITCHIE, J. B. 1995. Tourism: principles, practices, philosophies, John Wiley And Sons.

34. MURPHY, P., PRITCHARD, M. P. & SMITH, B. 2000. The destination product and its impact on traveller perceptions. Tourism management, 21, 43-52.

35. OLIVER, R. L. 1999. Whence consumer loyalty? *Journal of marketing*, 63.

36. OLIVER, R. L. & DESARBO, W. S. 1988. Response determinants in satisfaction judgments. *Journal of consumer research*, 495-507.

37. OPPERMANN, M. 2000. Tourism destination loyalty. Journal of travel research, 39, 78-84. PARK, D.-B. & NUNKOO, R. 2013. Relationship between Destination Image and Loyalty: Developing Cooperative Branding for Rural Destinations.

38. RAJESH, R. 2013. Impact of Tourist Perceptions, Destination Image and Tourist Satisfaction on Destination Loyalty: A Conceptual Model. PASOS. Revista de Turismo y Patrimonio Cultural, 11, 67-78.

39. RANASINGHE, J. TOURISTS'SATISFACTION AND BEHAVIORAL INTENTIONS: ANEMPIRICAL EVIDENCE FROM TOURISM INDUSTRY IN SRI LANKA. Proceedings of International Conference on Business Management, 2012.

40. ROODURMUN, J. D. J., T 2010. Influence of trust on destination loyalty – An empirical analysis

41. The discussion of the research approach. International Research Symposium in Service Management.

42. SIPE, L. J. 2013. Leadership for innovation in the memories business: A mixed methods study of a hospitality and tourism marketplace, University of San Diego.

43. SNEPENGER, D., MEGED, K., SNELLING, M. & WORRALL, K. 1990. Information Search Strategies By Destination-Naive Tourists. Journal of travel research, 29, 13-16.

44. SWEENEY, J. C. & SOUTAR, G. N. 2001. Consumer perceived value: the development of a multiple item scale. *Journal of retailing*, 77, 203-220.

45. UNWTO 2013. The World Tourism Organization (UNWTO) Tourism Highlights.

46. WOODRUFF, R. B. 1997. Customer value: the next source for competitive advantage. *Journal of the academy of marketing science*, 25, 139-153.

47. WOODSIDE, A. G. & MACDONALD, R. M. 1994. General system framework of customer choice processes of tourism services. Spoilt for choice, 30.

48. YANG, Z. & PETERSON, R. T. 2004. Customer perceived value, satisfaction, and loyalty: the role of switching costs. Psychology & Marketing, 21, 799-822.

49. YOU, S. 2011. Effects of Food Image on Tourists' Destination Image and Visit Intention. ŽABKAR, V., BRENČIČ, M. M. & DMITROVIĆ, T. 2010. Modelling perceived quality, visitor satisfaction and behavioural intentions at the destination level. Tourism management, 31, 537- 546.

11

FROM PRIM TO PUNK: THE SUCCESSFULLY ALTERED MARKETING IMAGE OF CHRISTCHURCH CITY POST-DISASTER

Linda Jean Kenix

Media and Communication, School of Social and Political Sciences, University of Canterbury, New Zealand.

ABSTRACT

Since a devastating earthquake in 2011, tourism marketers have emphasized events and organizations in Christchurch that highlight independence, creativity, innovation, participation, and art. This research examines how promotional materials for three specific organizations and events framed their own organizations as well as the city itself. This research suggests that a focus on collaborative creativity in tourism marketing may have resulted in a very clear competitive advantage for Christchurch as it rebuilds. Lonely Planet named Christchurch as one its top ten cities for its 2013 Best in Travel and The New York Times listed it as the 2nd of "52 Places to Go in 2014"- only a few scant years after an earthquake that destroyed almost the totality of the central city. This is an unlikely outcome that was only possible when the government, which itself is strongly interconnected with conglomerate corporations and sponsored networks of mainstream businesses, made a dramatic shift in marketing strategy toward a reinvention of Christchurch as a centre for alternative, independent artistry.

KEYWORDS

Tourism, space, media, Christchurch, earthquake, marketing, authenticity, participation

1. INTRODUCTION

On 22 February 2011, a devastating 6.3 magnitude earthquake struck Christchurch, New Zealand and killed 185 people. While the magnitude was not as large as many other, less devastating earthquakes, it struck at a shallow depth of 5 kilometres, within 10 kilometres of the city, and in the middle of a working day (Geonet 2011) – three factors that increased its impact dramatically. The earthquake itself lasted only seconds, however the physical and personal destruction was tremendous. In the years that followed the earthquake, the city experienced intense deconstruction. By the beginning of 2015, nearly all of the estimated 1,611 damaged properties on land that was 'green zoned,' meaning it could again be built upon, were fully or partially demolished (Stylianou 2013). Another 7,037 properties were 'red zoned' for complete removal. As of early-2015, 5,824 'red zoned' properties were cleared. The deconstruction of Christchurch occurred concomitantly with increasingly intensive construction of remediated or renewed physical spaces. Approximately 29,000 building permits were granted in Christchurch from 2011 until 2015. In the central city, 240 new buildings were under construction or completed by the end of 2014. This physical reengineering occurred alongside $2 billion in infrastructural work on roadways and bridges (One News 2015). In 2014 alone, approximately 90 thousand square meters of road was replaced and 294 kilometres of wastewater pipe repaired (Canterbury Earthquake Recovery Authority 2015). The total estimated cost of the earthquake at the beginning of 2015 was $40 billion and rising.

This radical revolution in physical space led to an equally radical shift in marketing how tourists could engage imaginatively and physically with the transforming city space. Prior to 2011, Christchurch was framed principally as a refined attraction with 'English' sensibilities, highlighting activities such as punting the Avon, theatre in Hagley Park, and riding the gondola. These activities were marketed as essential components of a Christchurch experience, defined as the most English city in New Zealand. If a tourist was to participate in that coded English space, they needed to engage with the attractions necessary for that experience as defined by marketers and other previous travellers. However, after 2011, the marketing of Christchurch shifted dramatically to highlight independence, collaboration, creativity, innovation, and art – and visitors were actively invited in to engage with these newly marked tourist spaces.

Specific tourism websites and government-supported events in Christchurch will be examined as components of a larger animated dialogue about space, power and place in the city. The commercially constructed invitation to tourists within what is coded as an engaged and performative space, has granted profound power to the individual visitor. This research will explore three heavily promoted organizations and events in Christchurch: the Spectrum Street Art Festival, the community organization Gap Filler, and Christchurch Art Gallery's Outer Spaces. All three of these organizations and events have been strongly supported by the Christchurch City Council and repeatedly featured in promotional materials as essential tourist attractions

This shift in marketing and event management in the city toward organizations and events such as these was driven by very obvious physical and spatial changes, but has resulted in what appears to be a competitive advantage for Christchurch. The result of this focused marketing directive is that Lonely Planet named Christchurch as one its top ten cities for its 2013 Best in Travel and The New York Times listed the city as the 2nd of "52 Places to Go in 2014"– all only a few scant years after an earthquake that destroyed almost the totality of the central city. This is an unlikely outcome that was possible when marketers reimagined the city as a space for alternative, independent artistry and bestowed power to the individual visitor to engage with those spaces as a collaborator.

2. LITERATURE REVIEW

2.1 The Power of Media Frames

Media inform audiences about places and construct the salience, perception, and memory of the world around us. In single moments, as well as over the course of time, media frames, "are manifested by the presence or absence of certain keywords, stock phrases, stereotyped images, sources of information, and sentences that provide thematically reinforcing clusters of facts or judgments" (Entman 1993: 52). This social construction of reality is understood as an invisible process involving media-generated words and images, which then form shared meanings about the social and political world (Gamson et al. 1992).

Carragee and Roefs (2004) argued that framing studies must begin to examine their results within the 'contexts of the distribution of political and social power' (214). They build upon previous research which broadly, yet directly, linked framing to power and ideology (Gitlin 1980; Tuchman 1978). While this relationship is dynamic and multi-directional, it is predominately the elite, powerful, media that shape how the public interprets the world (Sotirovic 2000), rather than the other way around. Consequently, the public's main understanding of places, people and events around the world are derived from a framed construction provided by media over time and from a select perspective (i.e. Gamson 1992; Gitlin 1980; Altheide 1976; Ryan et al. 1998; Tuchman 1978). This perspective generally favours elites and bestows power to those who frame the message, rather than those who engage with that message as audience members.

Gitlin (1980) long ago defined frames as "persistent patterns of cognition, interpretation, and presentation, of selection, emphasis, and exclusion, by which symbol-handlers routinely organise discourse" (7). A frame determines what is "relevant" (Hertog and McLeod 1995: 4) and "suggests what the issue is" (Tankard Jr. et al. 1991). In regards to tourism, the frame chosen by elite symbol-handlers determines the relevance of a selected event or organization and places that frame into larger narratives about what does and does not qualify as a worthy tourist space.

While there are manifest elements that comprise visual and textual frames, there are also latent sociological, political, and cultural cues embedded within messages (Huxford 2001; Kenix 2008) that all coalesce to expose the ideological constituency (Reeves and Campbell 1994; Manovich 2001; Kenix 2009) of those who created the media message. Creators of media content exist within a mutually inter-supported nexus that simultaneously reflects and perpetuates social contexts (Julier 2000; Barnard 2005). As such, media frames must not be understood as manifestly or purely evidentiary. Rather, frames (both textual and visual) are "organizing principles that are socially shared and persistent over time, that work symbolically to meaningfully structure the social world" (Reese et al. 2001: 11). This structure of meaning is fundamental to mediated representations of tourist spaces. An individual's understanding of place is often formed by mediated representations, or frames, of that physical space as well as mediated constructions of how and why people engage with that space (Jansson and Falkheimer 2006). Individuals searching for information about potential tourist locales likely have no direct experience with those regions before they plan their travels. They depend upon these framed representations to make crucial decisions about their travel plans. Economies rise and fall based on these representations. As the world "becomes a complex web of intertexts and hypertexts, reality, media and tourism are more and more closely intertwined" (Davin 2005: 178)

2.2 Mediating Tourist Spaces

Tourism is influenced by how media frame potential tourist spaces (Campbell 2005), which themselves can be seen as a metaphor. A tourist space is a metaphor for a geographic, mental or abstract space (Kowalczyk 2014) that only becomes a tourist space when marked as such by tourists or tourism marketers. Thus, how a tourist space is constructed and marketed by tourism marketers is crucial to the success of that space as a tourist attraction and of how individuals then think of that tourist space. This research attempts to examine how marketers may have framed Christchurch as a space for participatory and authentic creativity. The marketing of Christchurch as a space for innovation and collaboration may appear as intuitive given the contemporaneous nature of physical reconstruction occurring in the city, however this marketing decision, like all others, is one made amongst a range of competing options and therefore demands exploration.

Tourism is fundamentally "semiotic in nature" (Jaworski and Thurlow 2011: 363), informed by "mediatized" representations (Månsson 2011) that then inform how tourists engage in performative spaces, otherwise known as tourist attractions. The tourist participates within these semiotic exercises by perpetuating learned mediated representations, but also producing those same spatial experiences that were initially sought from those mediated moments (Kahn 2003). These defined spatial codes for tourists trap willing participants within their own 'hermeneutic circle' (Urry 2002) of predetermined expectations to engage with and travel through. Tourists move imaginatively and performatively with previously constructed ideations of 'authentic' tourist experiences, created from formal, professional communications such as television programmes and tourist websites as well as from informal shared sources, such as Facebook and Flickr. Then, when tourists fulfil their own expected performative engagement within these predetermined spaces, they often share their imagery, memories and stories through what has been called a tourist haze (Thurlow and Jaworski 2012). This haze is a perpetual construction and reconstruction of what it means to be a tourist in a particular space.

The tourist space, indeed any space, "is as much a social construction as it is a physical phenomenon" (Jaworski and Thurlow 2011: 363). Physical spaces are continually in a process of becoming as the interaction between ideas and spaces are negotiated. This process has been labelled an 'animated dialogue' (Kahn 2003), due to the intransitive movement between a network of contributing forces. This perpetual process of 'becoming,' resulting in an evolving lived space, concomitantly co-constructed between conceived and perceived spaces (Lefebvre 1991), is performatively obvious in the context of a post-disaster reconstruction. The distinctions in such a context, between 'tourist' spaces and 'non-tourist' spaces are inherently problematic. However, tourists, in general, are interested in "the spaces between the 'important' must see, the banal objects of the everyday (that) play a larger role in the creation of a sense of place than has been surmised" (Metro-Roland 2011: 40). The contemporaneous spaces in Christchurch, necessitated by constantly changing physical structures, have exposed the possibility of previously conceptualized 'banal objects' as spaces for potential attractions. Indeed, what can be labelled as 'tourists' space' is in many cases a result of the deliberate actions of companies offering tourism services" (Kowalczyk 2014: 14). However, the possibility of tourist spaces being marked as something categorically different in Christchurch emerged after the earthquakes of 2011 because so much physical space had changed. Spaces, of course, are always inherently contemporaneous, but the marketing of tourism in Christchurch after the earthquake exposed public space as conspicuously public and marked that space as explicitly participatory.

A space can only become available for tourism if it is marked as accessible (Wieckowski 2014). This hurdle of accessibility could potentially be difficult in a post-disaster context unless one considers the potentiality of allowing tourists to participate in the reconstruction of that damaged space. Participation has as much to do with authenticity as it does with engagement. Forced engagement, without authenticity, makes the participation seem contrived and not necessarily experienced. "An authentic experience...is one in which individuals feel themselves to b e in touch both with a 'real' world and with their 'real' selves" (Handler and Saxton 1988: 243). The importance is on the emphasis of feeling as the authenticity of the experience or object can actually be false or contrived, but it can't be framed, and therefore then believed, as contrived if the experience is to be deemed authentic. Engaging with that experience, and constructing media messages that stress that level of participation, greatly strengthens the experiential authenticity for tourists who combine their mediated exposure with their lived reality to meaningfully assess the expanse of a tourist space. A tourist can then activate what has been called existential authenticity (Wang 1999), when experientially engaging with and participating in authentic activities, rather than viewing objects as a bystander. This level of existential authenticity within a tourist space can only be achieved through engaged participation.

There are potential economic benefits to the community of a tourist space that engages visitors in existential authenticity. It is most likely because of these economic benefits that exploiting existential authenticity is a goal for many in the tourist industry. Disaster tourism is increasing globally because visiting these sites "draws people closer to the event" (Miller 2008: 118). The disaster becomes marked as a tourist space by marketers who attempt to bring potential visitors in "to experience the shock and awe in a voyeuristic reality-style manner" (Miller 2008: 118). Such a lived experience that individuals can participate within, instead of view from a distance, provides something "increasingly rare: a travel experience that is vividly, irrevocably authentic" (Potts 2006). This research aims to study how marketers explored and exploited that authenticity in selling Christchurch as a potential destination for tourists.

3. METHODOLOGY

3.1 Research Question

The websites of three heavily promoted organizations and events in Christchurch were examined for this study. This research examined the following research question: How do the Spectrum Street Art Festival, Gap Filler, and Outer Spaces explore and exploit participation and authenticity when framing Christchurch as a tourist space?

3.2 Framing Analysis

All online tourist materials were analysed in accordance with earlier framing studies (i.e. Entman 1993), which search for general themes after examining the complete text as a whole. This research follows the example of previous work that examines individual frames from different sources, which then combine to form an overarching narrative (Atkinson 2006; McGee 1990). This approach argues that meta-narratives are comprised of multiple frames and do not exist as an entirety, but rather as a multitude of pieces. Narrative itself is actually "a dense reconstruction of all the bits of discourse from which [they were] made" (McGee 1990: 70). These bits of narrative can culminate at a specific cultural site, such as a physical (Atkinson 2006) or conceptual space (Kenix 2010). A cultural site can be a mental recognition of meaning – of a Gestaltian whole that inevitably defines its parts. When one views individual copy of a travel website, or any narrative fragment that they have had previous experience with, they instantly recognize something within those pages that is not simply on the page. A broader meaning, which is comprised of previous narrative fragments, come together to constitute a cultural site that defines that particular travel website – and then inevitably, the space which the travel site defines.

Competing and conciliatory narratives that are conceptualised against one another to create a larger, comprehensive, narrative, create this broader meaning. These narratives actually work against and with each other to reformulate a new reality through what has been called an "ideological rhetorical force" (Chatman 1990: 198). Such force comes from the culmination of cultural value and identity. Storytellers build narratives within a dichotomous cultural framework that is shaped in relation to desires, values and motives (White 1984). This is done through polarizing language that pits "them" against a unified "us" who all relate along common cultural markers (Atkinson 2006). As a narrative forms through repeated frames, a common identity within a particular cultural sphere develops that is based on "the unity of experiences and actions" (Carr 1986: 149). This shared identity allows those within that participatory sphere to recognize common qualities within themselves, the text and each other. This shared cultural value and identity also inevitably leads to an increased understanding of what it is that is different in other cultural spheres (Kelly 2004).

This research examines how tourist sites in Christchurch are framed through a qualitative content analysis (Krippendorff 2004) which aims to explore the constitutive functions of cultural values and identity (see Atkinson 2006) by examining any depictions of shared and opposing cultural values and identity in content. When applicable, specific strategies of signification (Mitra and Cohen 1999) will be identified in content. As Foucault (1989; 1991) argued, and the work of Atkinson (2006) confirms, any narrative is inevitably not about a particular person, place or thing. Rather, narratives are part of a complex network of identity and power relations.

Thus, this methodological approach, drawing from the original work of Fairclough (1995) and the later work of Atton and Wickenden (2005), examines the interdependencies between textual, discursive and social practices through narrative fragments and constitutive functions. This critical approach also involves a systematic process and examination of lexical choices, questioning the range of possible vocabulary items that could have been utilized otherwise (Matheson 2005). These studies can be difficult to replicate and are quite labour intensive (Semetko and Valkenburg 2000). However, they also allow for a deeper and more complex level of analysis.

4. RESULTS AND DISCUSSION

4.1 Spectrum Street Art Festival

Longstanding and entrenched bureaucratic institutions, such as The Canterbury Community Trust, The Christchurch City Council, and Creative New Zealand, principally funded the rather radical Spectrum Street Art festival. These multi-million dollar organizations are not typically known for their expertise in underground graffiti art. However, even organizations with an expertise in graffiti art know the value of highlighting market leaders. Oi YOU!, the community organization that curated the festival, notes in their online promotional material, that the exhibition features "the largest Banksy collection in the Southern Hemisphere" (Streetart 2015a). This notation gives the exhibition a point of geographical distinction and relational prestige. Although Banksy still remains pseudonymous, his work is generally seen as the most widely recognized and most important graffiti art in the world. By connecting Oi YOU! with the largest collection of Banksy in the Southern Hemisphere, marketers are simultaneously equating this festival as important and worthy of visitation for tourists. Integrating Banksy into the festival suggests collaboration between common 'street art' and work that is widely celebrated. In building this connection, Oi YOU! demands an obvious demarcation of the Spectrum Street Art Festival as a celebrated tourist space.

While the draw of Banksy could likely be enough to bring visitors into the festival, it is the repeated power bestowed on visitors that is emphasized throughout promotional materials which emerges as the true attraction. In describing Spectrum, it is noted that "'Immersion' is the operative world for the festival, as visitors are invited to engage with the work, rather than view from afar behind a red velvet barrier" (Publica 2015). The didactic and oppositional framing of this 'engaged' work with that held far behind a 'red velvet barrier' connotes a space that visitors are openly welcomed to participate within rather than being excluded from, outside the confines of parameters that have been dictated by others. The use of 'red velvet' in this instance is mentioned in derision and intended to conjure up secluded, elitist spaces of which this festival is most certainly abdicating from

The entrance to the festival space is through a constructed 'thrift shop' with a sign that states "Enter Through the Thrift Shop." This conjures the work of Banksy, whose film, Exit Through the Gift Shop was nominated for an Academy Award for Best Documentary. However, by transforming the exit to the entrance and changing 'gift' to 'thrift,' the festival immediately demonstrates an attempt to problematize participant expectations and engage patrons in a game of subversion. This sense of play continues into a cordoned space filled with subtly, and not so subtly, subverted items that appear to be for sale. Individuals are initially under the impression that they must presumably work their way through the thrift shop to get to the 'real' exhibit (Figure 1). However, participants soon realize that they are already inside the exhibit – as part of the exhibit. Their job as patrons is to excise as much of the subversion from their surroundings as possible. The 'desk' at the 'shop' has a handwritten sign that says "be back in 5 minutes." In the absence of authority, participants are encouraged to examine their own expectations and then explore objects that contradict those perceptions

The thrift shop then feeds into a corridor that moves the individual up stairs to what is a contrived space replicating an R17 subway car in a New York subway station. However, because the individual is part of that space, moving through it and engaging with it as a participant and not an observer, the perception is one of authenticity. Much like the thrift shop, this space is meant to operate on several different levels of expectation and participation. Images of this space on the Spectrum website and on their Facebook page demonstrate the real, functional, and perceptual (Wieckowski 2014) components of this exhibit. What would presumably be expected as a stairway, which leads to 'appropriate' art, is suddenly the recreation of a disparate location. However, the intricacy of that recreation, matched with a lack of any official guidance from the exhibit organizers, invites introspection. The stained walls, graffiti covered car, smudged tiles, and off-centre subway doors that appear to be presumably in movement, all suggest a level of unexpected authenticity. The contrived location of this exhibit – in a stairwell – then invites the participant to expose that authenticity for the obfuscation that it is. However, this invitation is only made clear after experiencing the subterfuge of the thrift shop. The combination of these two spaces, gives the participant a full awareness of their role in this space, clearly marked as subversive.

Figure 1. Thrift Shop

The first Oi YOU! Event in Christchurch was the RISE festival held in late 2013 at the Canterbury Museum. It attracted over 248,000 visitors through the museum – more than any other event held there (Streetart 2015b). The festival extended outside of the museum to 15 walls across the city and several community events. George Shaw, the event organizer, said: "I think pre-quake, it would have been very difficult to produce a show like RISE. I think what happened with the quake...not only were the buildings levelled, but to one degree or another, people's mentality was levelled as well. People realized what they had to do was they had to do it for themselves....everything was gone. They literally cleared spaces and in those spaces created entertainment for themselves...people started to realize, this is cool... (Streetart 2015b)

Shaw is speaking directly to the active framing of this artistic and community enterprise as participatory. The extension of the RISE festival to external spaces throughout the city was both physical and ideological. The festival itself was framed as a Christchurch festival and not a festival in the Canterbury Museum. In doing so, there is both recognition of community participation that had already existed and also an explicit request for engagement from the visitor.

Figure 2. Subway

The 2015 Spectrum festival built up on this lexical reliance on immersion and took the exploration further through a series of treasure hunt clues and rewards all throughout the nearby Botanic Gardens that visitors and locals could find and then claim for works of art. This expansion of the festival's reach outside of any confined walls, invited those on the inside to venture out of doors. By doing so, the festival encouraged visitors to reconsider what counts as a marked space for tourism and for art. When participants brought tokens back to the festival and engaged with festival workers, they were awarded unique and individual art pieces. By using the art as prizes, Spectrum reduced the generally hierarchical space between patron and artist. The two were equally engaged with each other in a fundamental reframing of expectations.

This reshuffle of expectations was performative as well as ideological. The festival was aiming, not only to reconceptualise what constitutes tourism, but also what constitutes social structure within the artistic, political and social world of Christchurch. Shaw, the founder of Oi YOU! expands upon the ideological work necessary to reframe expectations throughout the many levels of society, when these may be working in an oppositional direction. "We need to be engaging with people. Making them realize that they are valued in society. The visual art world so long has been about exclusivity. Stuff that. We want to wrestle that back. We want to create environments where people enjoy moving through them. Where there is verve, where there is life, where there is interaction" (Streetart 2015b). Shaw's use of the world 'wrestle' in regards to the reclamation of art toward interaction is purposeful here to highlight that all ideological changes, including this one, are indeed a struggle. Power is never easily exchanged.

One can see power embedded in this messaging when compared against marketing information from other institutional museum spaces, such as the Smithsonian in Washington, D.C. The Smithsonian website is full of fascinating information about new exhibitions and details a plethora of interesting data about award-winning artefacts on display (Smithsonian 2015). There is a running list of events and exhibitions as well as tips on where to stay and eat in Washington D.C. There is also a predominant list of news stories, bulleted to the right of the page, which lists stories such as "Here's Why the Dutch Are So Tall" And "Who Was Mary Magdalene?" In all of this marketing material, The Smithsonian is talking to the viewer. The knowledgeable curators and researchers at the Smithsonian are telling the viewer what is important and why. This reflects a very different set of lexical choices that frames power and engagement quite differently than Spectrum in Christchurch. The solitary entreaty from the Smithsonian museum to engage with potential participants is in the form of a question, "Ask Smithsonian: Could the Volcano Beneath Yellowstone National Park Ever Erupt?" However, this too is not constructed as an equilateral exchange between sources of equal power. This is an invitation for the individual participant, as subservient to an entrenched system of power, to ask the far more knowledgeable institution for answers that only it can provide.

4.2 Gap Filler

Given the scale of demolition in Christchurch, there have been many resultant vacant spaces that have remained dormant. The spaces where buildings one stood have been reduced to levelled rubble. Gap Filler was created as a response by the community to fill these spaces with varying forms of creative artistry. Their initial efforts were presumably not crafted as tourist attractions, but rather, purely as a means to involve the community in rebuilding the city centre. The Gap Filler website notes that the group was created to give "Christchurch people an opportunity to contribute to their city's regeneration" (Gap Filler 2015a). However, the organization and its many works have evolved to be a central attraction for incoming visitors.

Gap Filler launched the Transitional City Audio Tour in late 2013, which gave individuals headsets to explore the central business district on foot. The project was intended for people "who were keen to explore and demonstrate their shared interest in the Transitional City, and a desire to critique the top-down, 'Master Plan'/Blueprint process" (Gap Filler 2015b). Visitors, as well as locals, were noted in Gap Filler's intended audience for this project given that "the city is now our gallery" (Gap Filler 2015b). This project stood as one of the first evidentiary turning points for Gap Filler toward a space that engaged tourists and locals across the entirety of a city in spaces previously marked as unused.

On the Gap Filler website, Coralie Winn, director of the organization, stated that Gap Filler brings "life, energy, positivity, creativity, to vacant sites in Christchurch" (Gap Filler 2015a). The group has found vacant, unused lots in Christchurch and installed photographic exhibits, held improvised dance performances, installed community chess boards, constructed cycle-power cinema screens, painted pianos as an invitation for play, built miniature golf challenges, held music concerts and poetry readings, constructed communal eating spaces, and much more. The organization has evolved to such a publicized point that the Christchurch Regional Tourist Organisation website, which is the central website for Christchurch tourism, now directly urges visitors to see various Gap Filler projects in the city. "The team at Gap Filler along with an army of volunteers, temporarily 'activate' vacant sites with creative projects. The result is a series of imaginative spaces that engage and entertain as they bring people together" (Christchurch City Council 2015). The term 'activate' is used frequently in descriptions of Gap Filler and by Gap Filler itself on their website, where the organization displays an enlarged map of Christchurch with flags of "gaps we've activated, or helped to activate!" (Gap Filler 2015a). In repeatedly using the term 'activate' in promotional materials, Gap Filler is suggesting a heightened level of energetic activity because of their presence; an exertion of power that would have otherwise not been evident on stagnant land. This transformation of energy and activity depends upon the participation and engagement from others – the 'we' in this instance, who combine to create a shared universe of collaborators. Interested individuals are strongly urged away from complacent viewership and into involvement. The use of an enlarged Christchurch map also serves to mark the available space for that participation in grand, expansive terms. The entire city and its surrounds are available for participation and 'activation.'

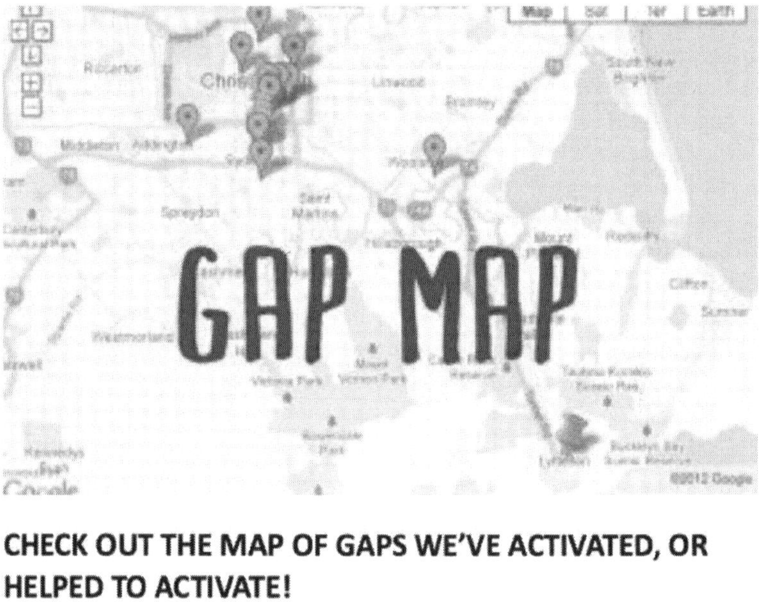

Figure 3. Gap Map

This need to 'activate' spaces is based on a narrative of empowerment. Dr. Ryan Reynolds, one of the co-founders of Gap Filler says that the organization "tries to empower the average citizen, to be able to transform the space of their city and their surroundings" (Gap Filler 2015a). A sense of empowerment comes from the self-actualized authorization of power – when an individual senses their own power in a way that may not been fully conceptualized beforehand and then recognizes potential 'activations' that could eventuate as a result. Those activations then transform not only the external spaces around Christchurch but also the internal spaces within each participant. Space becomes extrapolated both outward and inward in a continual process of 'becoming' for both the participant and the vacant physical space.

4.3 Outer Spaces

The Christchurch Art Gallery Te Puna o Waiwhetu has been closed since soon after the earthquakes. Although the building was used as Civil Defence headquarters following the massive earthquake, and was built specifically to manage seismic activity, it was found to require substantial remediation work. The gallery is funded by the Christchurch City Council and is noted as "an important part of the cultural life of this city and of New Zealand as a whole – an attraction to which you'll return again and again" (Christchurch Art Gallery 2015). Such a statement frames the gallery as a destination of importance. Indeed, the director Jenny Harper states plainly that "our collection is important and varied" (Christchurch Art Gallery 2015). She frames the work as necessary to be viewed for those who call Christchurch home and for those who are tourists. Given the obvious structural limitations of showing artwork in what was a closed gallery space, these important works moved to what the gallery has labelled 'outer spaces.' These outer spaces encompass all of Christchurch, as is evident in the imagery that the gallery used to promote their work. Like Gap Filler, Outer Spaces has relied on the map graphic to demonstrate the expanse of their reach.

The move outward may have been due to structural constraints, but obviously there were other choices available to the gallery. Simply renting a dedicated gallery space could have solved many of their temporary issues in viewership. However, the gallery chose to exploit that lack of temporal grounding and expose both the complexity and the simplicity in being fleet of foot. Director Harper states, "we enjoy providing artists the opportunity to create new works of art in our spacious foyer and galleries as well as outside the building" (Christchurch Art Gallery 2015). In this quote, the director exposes an obvious move toward creation and away from purely viewing as well as a movetoward the outside rather than inside. This transferral of space, from inward to outward, occurs alongside a transferral of action, from constricted to emancipatory. Massey eloquently writes that space "is about contemporaneity (rather than temporal convening), it is about openness (rather than inevitability) and it is also about relations, fractures, discontinuities, practices of engagement. And this intrinsic relationality of the spatial is not just a matter of lines on a map; it is a cartography of power" (Massey 2005: 85). The 'outer spaces' graphic demonstrates that cartography of power plainly – it belongs to everyone.

Figure 4. Outer Spaces

5. CONCLUSIONS AND RECOMMENDATIONS

The three organizations examined for this research - Spectrum, Gap Filler and Outer Spaces – all demonstrated a radical revision of what constituted public and tourist spaces. This revision was marketed to both tourists and locals as a reconceptualised imaginative and physical city space. Prior to 2011, Christchurch was framed principally as a refined attraction with 'English' sensibilities, highlighting activities such as punting the Avon, theatre in Hagley Park, and riding the gondola. However, after 2011, the marketing of Christchurch shifted dramatically to highlight independence, collaboration, creativity, innovation, and art – and tourists were actively invited in to engage with these spaces.

The dismantling of Christchurch's infrastructure led to a reconfigured spatial infrastructure. What was once closed became opened and what was once introspective in approach became inherently extroverted. This co-created process led to a larger animated dialogue about space, power and place in the city. The commercially constructed invitation to tourists within what was coded, marketed and 'sold' as an engaged and performative space, granted profound power to the individual visitor. The result of this ideological shift appears to have resulted in a competitive advantage for Christchurch, which was recently named by Lonely Planet and The New York Times as one of the best places to travel - only a few scant years after an earthquake that destroyed almost the totality of the central city. This is an unlikely outcome that was only possible when marketers reimagined the city as a space for alternative, independent artistry and bestowed power to the individual visitor to engage with those spaces as a collaborator.

This research has obvious limitations. One of the most pronounced is that only a very small sample of organizations was examined. However, even in this small sampling, some lessons were learned that may be useful for other countries, such as Haiti and Japan, as they struggle in their own post-disaster contexts to attract a tourist economy. The mutual construction of a collaborative and engaged space for visitors to co-create a space for empowerment, creativity and growth appears to have benefited tourists and citizens who wish to participate in the reconstruction of Christchurch as well as local organizations who are attempting to build physical and ideological relationships in a strained structural environment.

REFERENCES

1. Altheide, D. (1976) Creating reality: How TV distorts events, Beverly Hills: Sage.

2. Atkinson, J. (2006) "Analyzing resistance narratives at the North American Anarchist Gathering," in Journal of Communication Inquiry 30: pp. 251-272.

3. Atton, C. and Wickenden E. (2005) "Sourcing routines and representation in alternative journalism: A case study approach," in Journalism Studies 6: pp. 347-359.

4. Barnard, M. (2005) Graphic design as communication, London: Routledge.

5. Campbell, N. (2005) Producing America: Redefining Post-Tourism in the Global Media Age. In: Crouch D, Jackson R and Thompson F (eds) The Media and the Tourist Imagination: Converging Cultures. London: Routledge, pp. 198-214.

6. Canterbury Earthquake Recovery Authority. (2015) Greater Christchurch Recovery Update. (accessed 4 April 2015) http://cera.govt.nz.

7. Carr, D. (1986) Time, narrative, and history, Bloomington: Indiana University Press. Carragee, K. M. and Roefs W. (2004) "The neglect of power in recent framing research," in Journal of Communication 54: pp. 214-233.

8. Chatman, S. (1990) Coming to terms: The rhetoric of narrative in fiction and film, Ithaca: Cornell University Press.

9. Christchurch Art Gallery. (2015) Director's Welcome. (accessed 10 April 2015) http://christchurchartgallery.org.nz/about/.

10. Christchurch City Council. (2015) Things to do in and around Christchurch. (accessed 9 April 2015) http://www. christchurchnz. com/what-to-see-and-do/things-to-do-in-and-aroundchristchurch/.

11. Davin, S. (2005) Tourists and the television viewers: Some Similarities. In: Crouch D, Jackson R and Thompson F (eds) The Media and the Tourist Imagination: Converging Cultures. London: Routledge, pp. 170-182.

12. Entman, R. (1993) "Framing: Toward clarification of a fractured paradigm," in Journal of Communication 43: pp. 51-58.

13. Fairclough, N. (1995) Media discourse, London: Arnold.

14. Foucault, M. (1989) The archaeology of knowledge, London: Routledge.

15. (1991) Discipline and Punish: The birth of the prison, Harmondsworth: Penguin.

16. Gamson, W. A. (1992) Talking politics, New York: Cambridge University Press.

17. Gamson, W. A., Croteau D., Hoynes W., et al. (1992) "Media images and the social construction of reality," in Annual Review of Sociology 18: pp. 373-393.

18. Gap Filler. (2015a) Gap Filler. (accessed 10 April 2015) http://www.gapfiller.org.nz/.

19. (2015b) The Transitional City Audio Tour. (accessed 10 April 2015) http://www.gapfiller.org.nz/the-transitional-city-audio-tour/.

20. Geonet. (2011) Frequently Asked Questions About the Canterbury Earthquakes. (accessed 4 April 2015) http://info.geonet.org.nz/display/home/Canterbury+FAQ.

21. Gitlin, T. (1980) The whole world is watching: Mass media in the making and unmaking of the new left., Berkley: University of California Press.

22. Handler, R. and Saxton W. (1988) "Dissimulation: Reflexivity, Narrative, and the Quest for Authenticity in 'Living History'," in Cultural Anthropology 3: pp. 242-260.

23. Hertog, J. and McLeod D. (1995) "Anarchists wreak havoc in downtown Minneapolis: A multi-level study of media coverage of radical protest," in Journalism Monographs 151: pp. 1-48.

24. Huxford, J. (2001) "Beyond the referential: Uses of visual symbolism in the press," in Journalism: Theory, Practice and Criticism 2: pp. 45-71.

25. Jansson, A. and Falkheimer J. (2006) Towards a geography of communication. In: Jansson A and Falkheimer J (eds) Geographies of Communication: The Spatial Turn in Media Studies. Góteborg: Nordicom, pp. 9-25.

26. Jaworski, A. and Thurlow C. (2011) "Locating self: Embodiment and remediation in/of tourist spaces," in Visual Communication 10: pp. 349-366.

27. Julier, G. (2000) The culture of design, London: Sage.

28. Kahn, M. (2003) "Tahiti: The Ripples of a myth on the shores of the imagination," in History and Anthropology 14: pp. 307-326.

29. Kelly, J. (2004) Borrowed Identity, New York: Peter Lang.

30. Kenix, L. J. (2008) "From media frame to social change? A comparative analysis of same-sex rights in the United States and New Zealand Press," in Australian Journal of Communication 35: pp. 105-128.

31. (2009) "A historical framing analysis of liberalism in American print media (1978-2002)," in Southwestern Mass Communication Journal Fall: pp. 81-97.

32. (2010) "Resistance narratives in radical, alternative media: A historical examination of Earwig," in Equid Novi 31: pp. 89-113.

33. Kowalczyk, A. (2014) "The Phenomenology of Tourism Space," in Tourism Geographies 24: pp. 9-15.

34. Krippendorff, K. (2004) Content analysis: An introduction to its methodology, Thousand Oaks: Sage.

35. Lefebvre, H. (1991) The Production of Space, Oxford: Blackwell.

36. Manovich, L. (2001) The language of new media, Cambridge: MIT Press.

37. Månsson, M. (2011) "Mediatized Tourism," in Annals of Tourism Research 38: pp. 1634- 1652.

38. Massey, D. (2005) For Space, London: Sage.

39. Matheson, D. (2005) Media discourses: Analysing media texts, Berkshire: Open University Press.

40. McGee, M. (1990) In search of "the people": A rhetorical alternative. In: Lucaites J, Condit C and Caudill S (eds) Contemporary Rhetorical Theory. London: Guilford, pp. 341-357.

41. Metro-Roland, M. M. (2011) Tourists, Signs and the City: The Semiotics of Culture in an Urban Landscape, Farnham-Burlington: Ashgate Publishing.

42. Miller, D. S. (2008) "Disaster tourism and disaster landscape attractions after Hurricane Katrina," in International Journal of Culture, Tourism and Hospitality Research 2: pp. 115- 131.

43. Mitra, A. and Cohen E. (1999) Analyzing the web: Directions and challenges. In: Jones S (ed) Virtual culture: identity and communication in cybersociety. Thousand Oaks: Sage, pp. 55-79. One News. (2015) Christchurch counts the cost four years on from earthquake. Auckland: Television New Zealand, pp.

44. Potts, R. (2006) In New Orleans: The Allure of Diaster Tourism. (accessed May 7 2009) http:// travel.news.yahoo.com/b/rolf_potts/20060313/rolf_potts/rolf_potts2915.

45. Publica. (2015) PUBLICA. (accessed 10 April 2015) http:// publica.co.nz/news-spectrumlaunches-tonight/.

46. Reese, S. D., Gandy Jr. O. H. and Grant A. E. (2001) Framing Public Life, Mahwah: Lawrence Erlbaum.

47. Reeves, J. L. and Campbell R. (1994) Cracked coverage: Television news, the anti-cocaine crusade, and the Reagan legacy, Durham: Duke University Press.

48. Ryan, C., Carragee K. M. and Schwerner C. (1998) "Media, movements, and the quest for social justice," in Journal of Applied Communication Research 26: pp. 165-181.

49. Semetko, H. A. and Valkenburg P. M. (2000) "Framing European politics: A content analysis of press and television news," in Journal of Communication 50: pp. 93-109.

50. Smithsonian. (2015) Featured Smithsonian Events & Exhibitions. (accessed 11 April 2015) http://www.smithsonianmag.com/smithsonian-institution/events-exhibitions/.

51. Sotirovic, M. (2000) "Effects of media use on audience framing and support for welfare," in Mass Communication and Society 3: pp. 269-297.

52. Streetart. (2015a) Oi You! (accessed 6 April 2015) http://streetart.co.nz/.

53. (2015b) Oi You!: About. (accessed 6 April 2015) http://streetart.co.nz/about/.

54. Stylianou, G. (2013) 'Chch demolitions nearing end,' The Press, September 9, 2013, http://www.stuff.co.nz/the-press/business/the-rebuild/9140821/Chch-demolitions-nearing-end. Tankard Jr., J. W., Hendrickson L., Silberman J., et al. (1991) Media Frames: Approaches to Conceptualization and Measurement. Association for Education in Journalism and Mass Communication. Boston, pp.

55. Thurlow, C. and Jaworski A. (2012) "Elite Mobilities: The Semiotic Landscapes of Luxury and Privilege," in Social Semiotics 22: pp. 487-516.

56. Tuchman, G. (1978) Making news: A study in the construction of reality, New York: Free Press.

57. Urry, J. (2002) The Tourist Gaze, London: Sage.

58. Wang, N. (1999) "Rethinking authenticity in tourism experience," in Annals of Tourism Research 26: pp. 349-370.

59. White, J. B. (1984) When words lose their meaning: Constitutions and reconstitutions of language, character, and community, Chicago: University of Chicago Press.

60. Wieckowski, M. (2014) "Tourism Space: An attempt at a Fresh Look," in Tourism 24: pp. 17- 24.

INDEX

S

T

W